COSMIC SPACE IS GOD AND PHYSICAL UNIVERSE IS GOD'S DREAM

DR. CHANDRA BHAN GUPTA

B.Sc. (Lko.), M.B.B.S. (Lko.),
M.D. THESIS (MED). (ALLD).,
M.R.C.P.(UK),F.R.C.P. (Edin.), F.R.C.P.(Glasg.),
E.C.F.M.G. Certificate (U.S.A.)

With Love to My Children

Nirupama

Sujata

Anil

ACKNOWLEDGEMENTS

Let me firstly express my feelings of gratitude to all those who have been instrumental in helping me create this book.

Secondly, I am extremely thankful to my wife Diana, for her help in the preparation of the book despite her various other responsibilities, and to my daughter Sujata for her help. Thirdly, I extend my grateful thanks to my son-in-law, Sunil, whose invaluable technical and practical assistance was proffered notwithstanding his extremely busy schedule.

I would also like to mention many Adwaitic thinkers, scholars, teachers, seekers and students, who after reading my two previous books entitled – **'Adwaita Rahasya: Secrets of Creation Revealed'**and the follow up book entitled – **'Space is The Mind of God : A Scientific Explanation of God and His Abode'**, constantly urged me to write this book.

Finally, I express my sense of intense gratitude to Cosmic Space, who fills the whole cosmos, for choosing me as His instrument to pen this book.

Table of Contents

HAMMER BLOWS

The age-old ignorance of mankind with regards to the "**Basic Truth**" vis-à-vis the current universe is the "**Unyielding Wall**". Knowledge given in this book with regards to this "**Basic Truth**" is the "**Nail**" and the use of myriad synonyms with regards to each and every significant word which has been put in use in this book, vigorously and adventurously in order to give birth to the required amount of knowledge with regards to this "**Basic Truth**" are the "**Hammer Blows**" in order to "**Nail The Knowledge**" into the "**Unyielding Wall**" of "**Ignorance of Mankind**" with regards to this "**Basic Truth**".

Introduction

One of the ancient Knowledge Systems of India is designated as "Adwait – Vedanta". This ancient Knowledge System of India is the means of Discovering, Gathering, Obtaining or Securing the Empirical Evidence with regards to "The Source of the Physical Universe" and its "Prolegomenous Nature."

An Empirical Evidence is defined as the Evidence which is Verifiable by Experiments and Observations rather than by Mere Theory or Pure Logic.

This book by Dr. Chandra Bhan Gupta, entitled **"COSMIC SPACE IS GOD AND PHYSICAL UNIVERSE IS GOD'S DREAM,"** is based on this Ancient Knowledge System called "Adwait – Vedanta."

Knowledge provided in this book with regards to "The Source of the Physical Universe" and its "Prolegomenous Nature" is Empirically Verifiable by Experiments and Observations, but experiments and observations of the 'Consciousnessbal Kind" and not the experiments and observations of the "Physical

Kind." This is the case on account of the fact that Cosmic Space is an amazing "Non-Physical Truth" and not an ordinary "Physical Truth." Therefore, the knowledge provided in this book with regards to "The Source of the Physical Universe" and its "Prolegomenous Nature" is not a Mere Theory or Pure Logic because it can be Empirically Verified by Experiments and Observations of the "Consciousnessbal Kind." However, it cannot be Empirically Verified by Experiments and Observations of the "Physical Kind" for the simple reason that Cosmic Space is an amazing "Non-Physical Truth and not an ordinary "Physical Truth."

MAN'S SEARCH FOR UNITY

The human family of mankind has been kept divided for far too long on the basis of their outer characteristics. Some of these characteristics, which are employed by humanity time and again to divide them into many warring factions, are enumerated below:-

Race

Nationality

Skin color

Language

Religion

Food and dress habit

Socio-economic condition

Geography

Each of these divisions is further fragmented along the sub-traits of countless other varieties. These sub-traits, even though they have been labelled here as sub-traits, they all, without exception, are of surface kind only and not of the fundamental kind.

Notwithstanding the superficial nature of all the above -mentioned traits or characteristics, they, nonetheless, have become the cause of more and more splintering or fragmentation of one human tribe or clan, into a large number of sub-tribes or sub-clans and beyond on account of mankind's nescience or ignorance with regards to mankind's fundamental unity.

These sub-tribes or sub-clans, seem most of the time at odds with each other, and many a time even "at dagger's drawn" at each other, totally forgetting or ignoring the fact that they all belong to one and only one race or one and only one species called the human race or human species and they all, without exception, have issued forth or have come out of the womb or more precisely, have issued forth or have come out of the awareness of the one and only incredible god or creator of the present-day cosmos.

Thus, the above -mentioned division of one human family, first into many races and then into large number of sub-races or sub-groups, on the basis of external differences, leads them to think that they are intrinsically or fundamentally different from one another. This, in its turn, becomes the reason for some groups to develop an illusionary sense of fundamental superiority over the others. This imaginary or false notion of innate superiority of some groups of human beings over the others forms the basis of never- ending suspicion, dislike, fear, hatred and finally war and domination of some human sibs over the others.

This breeds further suspicion, dislike, fear and hatred and not surprisingly, the desire for revenge in the awareness' of those who find themselves at the receiving end of suppression and domination.

The oppressors, on the other hand, feel vindicated, albeit erroneously, in their sense or feeling of superiority or pre-eminence over those upon whom they have managed to prevail by use of force or by some other equally detestable means. Thus, a vicious or savage, never ending cycle of hatred and counter hatred, revenge and counter revenge plus violence and counter violence of increasing ferocity becomes deeply ingrained in the psyche or the awareness' of these opposing groups, leading to, in many instances, what is described as a 'thousand year war' or a 'thousand year confrontation', generation after generation despite the fact that they all are, they all were and they all will always be the sibs or siblings because they all have issued forth or they all have come out of the womb or , more precisely, they all have issued forth or come out of the awareness of one and only, incredible creator or god who is the creator or god not only of them but of the entire present-day space time and physical universe.

The loud message to entire mankind is :-

The above described, very sad and disappointing behaviour of their's is on account of their ignorance of their common origin from one and one only Creator or God of the universe who is invisible to the naked eye but it nevertheless, is present or immanent, all the while, behind, before, beside and even inside each and every member of the human family, nay, each and every denizen or inhabitant or occupant or being of the entire present-day cosmos.

ADWAITISM

The magnificent Adwait-Vedanta pronounces loud and clear to mankind at large that BRAHMAN aka God, The Source, or The Fountainhead of the current physical as well as non-physical universe of mankind's fully-awake or wide-awake-state is one and one only and not two, three, four, five or whatever. However, this Majestic BRAHMAN aka God, The Source or The Fountainhead has as many manifestations as there are BEINGS and THINGS in the concrete current cosmos, objective current cosmos or physical current cosmos and each and every such BEING and THING of the concrete current cosmos, objective current cosmos or physical current cosmos is one hundred percent authentic or genuine and inseparable or indivisible part of this one and only Majestic

BRAHMAN aka God, The Source or The Fountainhead because they all, without exception, are a mere daydream or oneiric of this Majestic BRAHMAN aka God and thus, one day, they all, without exception, will melt back, meld back, merge back, unite back, fuse back, coalesce back or dissolve back into this Majestic Brahman aka God, The Source or The Fountainhead in the manner all daydreams or oneirics are destined to melt back, meld back, merge back, unite back, fuse back, coalesce back or dissolve back into their daydreamer or oneirickizer.

The current universe of mankind's fully-awake or wide-awake-state which consists of both physical as well as non-physical truths or realities, is a mere daydream or oneiric of this Majestic Brahman aka God, The Source or The Fountainhead and thus, one day, is destined to melt back, meld back, merge back, unite back, fuse back, coalesce back or dissolve back into this Majestic Brahman aka God, The Source or The Fountainhead whose daydream or oneiric it is. The daydream or oneiric and its daydreamer or oneirickizer, even though at the perceptual and experiential level, seem or appear two separate truths or realities, but at the fundamental level, they are **NOT TWO** but **ONE** and **ONE ONLY**. This is the unique message of **ADWAITISM**, the heart of **Adwait-Vedantic Cosmology**.

The above mentioned, stunning or amazing **INSIGHT** offered by the magnificent Adwait-Vedanta to mankind at large vis-a-vis or in regards to the current universe of mankind's fully-awake or wide-awake-state which has both physical as well as non-physical dimensions or horizons, must not be called POLYTHEISM. Adwait-Vedanta is ADWAITISM. It is neither POLYTHEISM nor PAGANISM nor is it MONOTHEISM or ATHEISM. It stands ALONE amongst all the **"ISMS"** on offer to mankind till to date.

From what has been said above, one must clearly grasp that Adwait-Vedanta is ADWAITISM and ADWAITISM only and nothing but ADWAITISM and thus, it is neither Monotheism in the way of all Abrahamic Faiths, nor is it Polytheism in the way of Paganism. And of course, it is not Atheism. This is the brilliance and uniqueness of the offering of Adwait-Vedanta to mankind as a whole.

~*~*~*~*~

PHYSICAL, AWARENESSBAL AND SUPRA-AWARENESSBAL DIMENSIONS OF MAN-1

Man has three dimensions, namely :-

Physical.

Awarenessbal.

Supra - Awarenessbal.

1. Man's Physical Dimension

This dimension of man embraces or comprises the physical body of man. And physical body of man is one and only component of man's three-dimensional existence which is mortal or ephemeral or is subject to death.

The other two are not.

In the absence of its resident awareness, man's physical body is absolutely insensate or inanimate or azoic or mineral because, by itself or in itself, it intrinsically is insensate or inanimate or azoic or mineral.

In other words, man's physical body, in itself or by itself, is not inherently self-illumined or self-aware or self-conscious or self-sentient or self-sensate or self-spirited in the manner man's awareness or consciousness or sentience or mind or spirit is.

2. Awarenessbal Dimension of man

Man's awarenessbal dimension encompasses man's awareness which is also referred to as man's consciousness or sentience or mind or spirit or "I".

Man's awareness enables his insensate physical body to become alert. It also endows man's insensate physical body, the power of appreciation, comprehension, cognizance and discernment. It empowers the body with the ability to experience and imagine. It affords the body, the ability of lucidity and objectivity. It grants the body, the power of sentience. It is the source of body's thinking and understanding ability. It is the fountain-head of man's emotions. It is the magnificent entity in man's body which addresses itself as 'I'. This awareness in man's body which calls itself 'I', is the source or the author of two very important proclamations emanating from man's body namely :-

"I exist".

"I am an aware being or a conscious being or a sentient being or a mindful being or a spiritual being in absolute contrast to a brick or a stone, both of which are not".

In the absence of its awareness, man's body becomes totally "senseless to the world" or totally "unresponsive or nonreactive to the world", even if it is still alive and warm with a beating heart a pair of breathing lungs. Such a situation is a daily occurrence when man's awareness temporarily bids 'good-bye' to the body on the descent of man into his deep sleep state at night or any other time. It also occurs when man's awareness comes under the influence of general anesthesia or some other central nervous system depressant chemical. It also happens when man's physical body becomes afflicted with some serious metabolic illness such as diabetic coma or uremic coma or some serious neurological illness such as a massive stroke.

Man's "awarenessbal dimension" is also called man's

"consciousnessbal dimension" or man's "sentiencel-dimension" or man's "mental dimension" or man's "spiritual dimension".

3. Man's Supra - Awarenessbal Dimension

Supra - awarenessbal dimension of man is the unseen source or author or creator of the universe. It is timeless.

This unseen, "supra-awarenessbal dimension" of man has a genuine free will whereas man's awareness i.e. man's "awarenessbal dimension" has no such uncontested or undoubted free will. What is claimed by man's awareness as its free will, in truth, is merely a deluded thought on its part, nothing more nothing less.

The above statement can be put in a different way.

Man's awareness does not possess a genuine or authentic free will in the manner of its source or author or creator. What man's awareness innerly or internally feels as its individual or personal 'free will' is merely an apparent or seeming free will or merely a "mirage of the genuine free will " or better still, is merely an "inner mirage" which is induced or engendered into its thought process by the will of its source or creator.

In other words, idea of human awareness that it

possesses a free will, is   instigated or kindled into it by its source or author in order to bestow upon it a seeming feeling of having a free will but, in reality, this free will is one hundred percent specious or false or illusory because it is one hundred percent controlled by the absolute will of its source or author or creator namely god or whatever.

"Supra-awarenessbal dimension" of man is the unseen source or author or creator of man's other two dimensions namely the physical and the awarenessbal dimensions. Existence, survival and sustainability of latter two dimensions of man are entirely dependent on the absolute will of man's supra-spiritual dimension namely god or whatever.

Whether one believes it or not or accepts it or not, physical and awarenessbal dimensions of man are mere imageries or dreamries of god's awareness or mind. They both have been engendered or produced by god or the source or the author of universe inside its own awareness or mind through the very well- known or by the very common 'o' garden activity called daydreaming or imagery-making or dreamry-making or idle-fancy-making, no more no less.

The above statement can be put in another way.

Universe is a make-believe creation of god's awareness or mind because it is a product of god's awarenessbal-imagining or mental-imagining.

It is not just the physical body and awareness of man which are imageries or dreamries of god' awareness or mind but the entire space time and physical universe too are mere imageries or dreamries of god's awareness or mind, formed or produced by the latter inside its own awareness or mind through the process of daydreaming on its part or through the instrumentality of awarenessbal imagery-making on its part.

Since man's physical body and awareness are both products of daydreaming on god's part or, if it is preferred, are both mere imageries or dreamries of god's awareness or mind, forged or fabricated by god inside its own awareness or mind through the instrumentality of daydreaming on its part or through the instrumentality of awarenessbal imagery-making or dreamry-making on its part, it is not surprising that they both exist and flourish as long as god, god who is their creator, wants them to exist and flourish , no more no less .

The net gain for man, after the acquisition of the above knowledge, will be the realization on his part that even after the inevitable death of his physical body, his awareness or his spiritual dimension will continue to exist in its supra-awarenessbal form or in its supra-spiritual form.

At his absolute level i.e. at the level of his supra-awarenessbal or supra-spiritual dimension, man is immortal.

If man allows himself to wallow in ignorance that his existence or reality is limited or confined only to his time bound or mortal physical body then inevitably he will suffer mental anguish & consternation throughout life with regards to his approaching, albeit ostensible, death at the ceasing of life in his physical body.

For man to view his total reality purely in the terms of or purely from the perspective of his perishable physical body is unpardonable. This wrong thinking on his part leads him on to the path of totally undeserved life-long mental angst and foreboding. Therefore, he must learn to educate himself that apart from his body, he also is the proud owner or possessor of his second dimension as well as an unseen third dimension both of which are eternal or immortal or timeless.

~*~*~*~*~

PHYSICAL, AWARENESSBAL AND SUPRA -AWARENESSBAL DIMENSIONS OF MAN -2

The strong sense of possessing a personal free will (to do whatever one likes), of which every human awareness is in the know of within itself, has been deeply imprinted in to its inner self by its source or the author with a view to make it remain anchored to the following two erroneous notions :-

That the physical universe including the physical body of man is one hundred percent authentic or genuine, even though the absolute truth about the physical universe is quite different which is that it merely is an awarenessbal imagery or dreamry or fantasy or bubble or chimera of god's awareness and nothing more substantial than that, irrespective of whether human awareness believes this truth or not or accept this truth or not or not. By the way, physical universe can only be enjoyed in ignorance (of its above described true nature).

In order that human awareness holds on to its above-mentioned viewpoint steadfastly, its source or the author, has imprinted another notion into its inner self so as to strengthen further its first, above described erroneous viewpoint.   And this second erroneous notion of human awareness consists of its strong inner feeling that it possesses a personal free will to do whatever it likes.

Since   the absolute truth is that the present-day physical universe is merely an awarenessbal imagery or dreamry or fantasy or bubble or chimera of god's awareness, forged or fabricated by it through the instrumentality of daydreaming on its part , how can it ever be possible for any being of this daydream-stuff composed universe, man or otherwise,   to have a genuine free will,   a genuine free will which is independent of the will of its source or the author or the creator namely god or whatever.

From time immemorial, physical universe has been an unsolvable riddle or enigma or puzzle for humanity to solve or decode or decipher. This has been the case simply because for a very longtime humanity has been pursuing a wrong track. This wrong track consists of the fact that mankind erroneously thinks and believes that the physical universe is absolutely authentic or genuine or real. But the supreme truth about the physical universe is that it is nothing of the sort. Instead, it merely is a product of god's daydreaming and hence it is composed of god's daydream-stuff and nothing else. In other words, the warp and woof or the essence or the make-up of present-day universe is "awarenessbal" and not something naively called "physical" by mankind.

The above can be put in another way.

Physical universe is merely a "make-believe-creation" of god's awareness or mind and thus it is made up of or composed of, or built of god's mind-stuff or awarenessbal-stuff and nothing else.

By the way, god itself is an incredible timeless, dimensionless awareness or mind of infinite intelligence.

It is absolutely absurd for any human awareness to think that it possesses a genuine free will, a genuine free will which is independent of the will of its source or author or creator namely god or whatever. To think in this tone or manner or vein on the part of any human awareness is totally illogical.

The supreme truth is that human body and awareness are both under the exclusive stewardship of god's will.

And so human awareness has no genuine free will, a genuine free will which is independent of the will of its source or author or creator namely god or whatever.

To reiterate.

The deep feeling which every human awareness harbors inside itself that it possesses a genuine free will, a genuine free will which is independent of the will of its source or author or creator called god or whatever, is merely a delusion on its part and nothing but a delusion on its part, a delusion which has been deliberately inserted or embedded into its inner self by no less a being than its source or author or creator in order to implant a misapprehension or an error into its inner self that the present-day universe is one hundred percent physical, that is to say, the present-universe is one hundred percent hard, solid, real and substantial or material and not merely or simply an "awarenessbal construct" or merely or simply an "awarenessbal assemble" or better still, and not merely a phantom or fantasy or phenomenon or dream or dreamry or imagery of god's mind or awareness.

Dr. Chandra Bhan Gupta

~*~*~*~*

PHYSICAL, AWARENESSBAL AND SUPRA - AWARENESSBAL DIMENSIONS OF MAN -3

Man must attempt to rise above his two obvious or easily discernible dimensions called his body and awareness or body and mind. While doing so, he must, additionally, try his utmost to accord recognition to his not so easily discernible or detectable supra-awarenessbal dimension who truly is the source or the author or the creator of his body and awareness or body and mind, irrespective of whether man believes this truth or not or accepts this truth or not.

To repeat.

Man must do his maximum to own up to his not so approachable and frustratingly imperceptible supra-awarenessbal dimension called god or whatever. Do this he must if he wishes to go beyond his mundane or humdrum or dull or boring or tiresome temporal existence. However, he must not undertake this task with a notion or feeling or sense of externally enforced compulsion or with a sense of grudge or with a sense that he is having to chase someone who is alien to him or he is having to say 'hi' or 'hello' to someone who is totally stranger to him.

Man must also not embark on this sojourn of seeking his third dimension with the feeling as if he is going to meet someone who dwells in some absolutely unknown, far away and bizarre place called heaven or paradise or someone who is entirely separate and different from his two common-'o'-garden dimensions namely body and mind or body and awareness or someone who is controlling and manipulating his body and mind or body and awareness as toy or puppet or plaything or sport. Neither he should undertake this search as if he is indulging in some kind of wild-goose chase.

Instead, he must seek the source or the author or the creator of his two dimensions in question in a manner as if he is in the quest of someone who truly and absolutely is the most important, most intimate and most enchanting or captivating part of his being and who dwells, not somewhere outside him, but inside him as a constant companion or as a bosom friend or as a sincere guide whose voice inside all mankind is called 'conscience'.

Until man finds the source or the author or the creator of his earlier mentioned two dimensions within his own truth or reality, he will not be able to free himself from the age-old delusion of possessing a genuine free will. And unless he is successful in uprooting this delusion of his, he will continue to regard himself a mere mortal being.

It is the delusion of possessing a genuine free will which is the root cause of man's all worldly suffering. This delusion is akin to a "noose" around the neck of humanity. This "noose" condemns humanity to the erroneous idea of man's mortality. Therefore, it is the foremost duty of man to free himself from this "noose" namely the false notion of possessing an individual free will which is independent of the will of his source or the author or the creator called god or whatever.

PHYSICAL, AWARENESSBAL AND SUPRA AWARENESSBAL KNOWLEDGE

Just as each and every human being has three dimensions to his existence, similarly there are three kinds of knowledge relating to all three of these dimensions.

Man, therefore, must attempt to master all three of these field of knowledge, if he aspires to make best use of his unique human essence or quintessence.

The above mentioned, three varieties of knowledge, available to mankind for partaking or imbibing are :-

1. Physical knowledge.

2. Awarenessbal knowledge.

3. Supra - awarenessbal knowledge.

As said earlier, humanity must make all effort to acquire all these three kinds of knowledge if it is desirous for a really fulfilled life.

Those individuals who become the master of all the above three varieties of knowledge are variously described as the "apprehender or knower of the composite wisdom of creation or composite science of creation or composite craft of creation".

1. *Physical Knowledge*

Most of mankind's effort, time and treasure is spent in learning and tutoring only this branch of '*composite knowledge*' which is extremely unfortunate as far as the true purpose of unique human existence is concerned.

This very restricted interpretation of the term '*knowledge*' by humanity leads to its very truncated development.

Physical knowledge encompasses not only the usual gathering of information about the day to day happenings and functioning of the physical world but also of inventing, discovering, learning plus accumulating all kinds of scientific, technical, medical, managerial, entrepreneurial as well as artistic and the remaining worldly knowledge which are ever eager to entice humanity for its indulgence in them.

There is nothing wrong as such with the accumulation of physical knowledge by anyone. In fact, acquisition of physical knowledge is essential if an individual, desires a balanced growth of his unique human life.

However, what one will like to highlight specially here though is with regards to or vis-a-vis man's interest and involvement exclusively or only with the acquisition of physical or worldly knowledge and very meager or very little or, even worse, absolutely nil interest and involvement with the attainment of the other two wings or branches of the 'composite knowledge', namely the awarenessbal knowledge and supra-awarenessbal knowledge.

This kind of short sightedness on the part of human beings leads to an unbalanced development of their lives.

Therefore, human beings must endeavour to acquire all the three varieties of the composite knowledge, namely the physical, the awarenessbal and the supra-awarenessbal, if they aspire for a genuinely satisfying and fulfilling life

2. *Awarenessbal knowledge*

Awarenessbal knowledge consists of knowing the true nature of human awareness in the cosmos as well as the true purpose of human awareness in the cosmos.

This knowledge deals with the intimate relationship that exists   between the human awareness of the cosmos on one hand and physical matter of the cosmos on the other.

Physical matter of the cosmos includes the physical body of man.

Another aspect of the present-day cosmos which the awarenessbal knowledge tackles is the intimate relationship that obtains between the human awareness of the cosmos on one hand and the cosmic space of the cosmos on the other.

Furthermore, this awarenessbal knowledge constantly discloses or reveals plus highlights or illumines the pathway human awareness must choose and steadfastly follow if it is desirous of its ultimate unification with its source or the author or the creator namely god or whatever.

3. *Supra-awarenessbal Knowledge*

Supra - awarenessbal knowledge is the ultimate wisdom which leads to irrevocable unification of human awareness to its source or the author or the creator namely god or whatever.

This knowledge i.e. the supra - awarenessbal knowledge   is the one and only eternal knowledge there is anywhere which every human awareness must make all effort to acquire. And once it is successful in acquiring it, it then must hold on to it for good, if it is desirous of finding out the secret of its own immutability or immortality.

The physical knowledge of this cosmos, in stark contrast, is all very ephemeral or non-eternal.

THE SOURCE OF THE UNIVERSE -1

Those faiths of mankind which subscribe to the idea of an aware and supremely intelligent creator of the universe, they all have their own particular concept of this creator which is exclusive to them. These concepts may or may not tally with those of other faiths.

If anyone is interested in knowing all these different notions of the creator of universe, held by different faiths, one may read the scriptures of these faiths.

What one is proposing to do here is to provide an understanding of the creator of universe from the *'Adwaitic'* or *'Adwait-Vedantic'* or *'Non-Dualistic'* standpoint. This 'Adwaitic' or 'Adwait-vedantic' or 'Non-Dualistic' standpoint is also sometimes called the standpoint of *'Ekatva'* or *'Unity'* or *'Singularity'* or 'One and Only'.

'Ekatva' is a Sanskrit word which means 'One and Only'.

Those who acquire the understanding of the source or the author or the creator of universe from the 'Adwaitic' or the 'Adwait-Vedantic' or the 'Non-Dualistic' or the 'Ekatvic' or the 'Unity' or the 'Singularity' or the 'One and Only' standpoint are said to be in 'Adwaitic State' or 'Adwait-Vedantic State' or 'Non-Dualistic State' or 'Ekatvic State' or 'Unity State' or 'Singularity State' or 'One Only State'.

"Ekatvic State" or "Singularity State" or "One Only State" or "Unity State" or "Adwaitic State" or "Adwait-Vedantic State" or "Non-Dualistic State" is also referred to as "Param State" or "Supreme State".

"Param" is a Sanskrit word which means "Supreme".

There is another Sanskrit word which one is going to employ later while discussing about "The source of universe" and this word is "Gyan". The word "Gyan" means "knowledge". The phrase "Param Gyan" implies "Supreme Knowledge" or "Supra-Awarenessbal Knowledge" about the incredible, timeless, dimensionless and disembodied or disbodied plus supremely intelligent source or author or creator of the universe.

~*~*~*~*~

THE SOURCE OF THE UNIVERSE -2

'Adwaita' or "Adwait-Vedanta" is a "Supra-Awarenessbal knowledge" or better still, it is a "Supra-Awarenessbal Science".

It should be classed as a "Supra-Awarenessbal science" because with regards to the precision of its methodology vis-a-vis the exploration into the "Supra-Awarenessbal Realm" of the cosmos, it is absolutely brilliant.

However, one must add here a word of caution with regards to 'Adwaita'. And it is the following :-

"Supra-Awarenessbal Science of Adwaita" is a

precise science but precise not in the "Physical-Empirical" sense which is the basis of physical science. Instead, it is a precise science in the "Awarenessbal-Empirical" sense. "Awarenessbal-Empiricism" is the 24-karat gold standard or yardstick for measuring the accuracy of the findings of the unique "Supra-Awarenessbal Science of Adwaita".

'*Adwaita*' has its foundation deeply embedded in the vast amount of most precise or unerring "awarenessbal-observations". The whole edifice or structure of 'Adwaita' is supported by the most critical and meticulous intellectual plus intuitive analyses of these "awarenessbal-observations" which form the fundamentals or the first principles of '*Adwaita*'.

These "awarenessbal-observations" of "Adwaita" are within easy reach of each and every human awareness on daily basis when it goes to sleep each night, provided it is sufficiently interested in not only observing them but also in applying its formidable penetrative power to unravel the extraordinary meaning which lay hidden deep down in them.

THE SOURCE OF THE UNIVERSE -3

The "Supra-Awarenessbal Science of   Adwaita" states that the source or the author or the creator of the universe is an incredible timeless, dimensionless plus bodiless or disembodied awareness of supreme intelligence.

The "Supra-Awarenessbal Science of Adwaita" further states that the awareness of man which lives or resides inside a physical matter composed body and as a consequence is called an "embodied" awareness, is fundamentally not different from the incredible timeless, dimensionless plus bodiless or disembodied awareness of its source or the author or the creator as far as its intrinsic qualities are concerned.

In other words, " Supra-Awarenessbal Science of Adwaita" announces that, both kinds of awareness' i.e. the disembodied awareness of the source or the author or the creator of the universe on one hand & the embodied awareness of man on the other, possess or have the same intrinsic qualities and thus they both have exactly the same intrinsic nature, despite the fact that one amongst them is disembodied and the other amongst them is embodied. Therefore, to regard them as being fundamentally different or dissimilar from each other on the basis of their embodied-ness and disembodied-ness is a sign of gross ignorance on the part of humanity.

The kinship that exists between the disembodied awareness of the source or the author or the creator of the universe on one hand and the embodied awareness of man on the other is akin to the kinship which obtains between a formless ocean on one hand and a drop of that ocean on the other.

The formless ocean is an extremely large and an extremely long-lasting truth, while a drop of the ocean is an extremely tiny and an extremely transient truth.

However, the point to take note of here is that even though a drop of the ocean is an extremely tiny and an extremely transient truth as compared to its source namely the ocean, the fact remains that the drop of the ocean is extremely tiny and extremely transient in merely a very restricted or circumscribed sense and not in any kind of fundamental or primordial sense. This is so because the drop of the ocean will soon lose its individual identity of a mere drop under the onslaught of the hot rays of the sun and will become a wisp of steam and cloud which ultimately will turn once again into a drop but this time into a raindrop. The latter namely the raindrop, when it will ultimately fall into the ocean - which one day it must - it will once again become one with the ocean, or if it is preferred, it will become once again the same as the ocean or identical with the ocean.

For all that, during the above mentioned " identity-voyage " of the main actor in the above drama namely the "water", first as the vast, formless ocean, then as an ocean drop, subsequently as a wisp of steam and cloud, and finally as a raindrop, the original ocean drop maintains its innate nature of being nothing but the molecules of water, each of which, in its turn, maintains its own innate nature of being two atoms of hydrogen and one atom of oxygen.

Two atoms of hydrogen and one atom of oxygen constitute each and every molecule of the ocean too from which the ocean drop was cleaved off through the mighty force of the sun in order that the ocean drop could execute its own individual role in the overall scheme of the cosmos.

To sum up.

The ocean and its drop, notwithstanding the obvious differences between them, are both mere atoms of hydrogen and oxygen only and nothing else, compounded in the ratio of two to one. In other words, all said and done, this is the primal nature of both.

Of course, as hinted above, there are vast differences between the ocean and its drop with regards to their respective size, depth and power. As a result, ocean can sink a battle ship in an instant whereas its drop can do no such thing. Nevertheless, the fact remains that they both are fundamentally mere molecules of water and nothing else, each molecule containing two atoms of hydrogen and one atom of oxygen only and nothing else.

Similarly, a drop of god's awareness during its residence inside the physical matter composed human body, seems as if it is transient just like the physical matter composed human body plus afflicted by all kinds of limitations and constraints. However, while in residence in the human body, human awareness never loses its innate nature which is the same as the nature of its source namely god.

At the instant of death of the body, the drop of god's awareness which was resident in the body, instantly merges back into the ubiquitous and the infinite ocean of god's awareness to become one with god's awareness once again.

THE ENTITY IN MAN WHO ADDRESSES ITSELF AS 'I' - 1

The entity in human beings who addresses itself as 'I' is human awareness.

This entity 'I' is the only portal or conduit there is for human beings to know or access the entity called universe and its source or the author or the creator called god or whatever.

Without this 'I' or without this human awareness, the existence of the universe and the existence of its source called god or whatever will both remain inaccessible or unknown forever as far as human beings are concerned.

Let's explore what one means when one says :-

"Without this 'I' or without this human awareness, the existence of the universe and the existence of its source called god or whatever will both remain inaccessible or unknown forever, as far as human beings are concerned".

It means that even if the entity called the universe and its source or the author called god or whatever are both one hundred percent believed as well as accepted by the human "I" or the human awareness as being capable of existing in the eternal absence of the human "I" or the human awareness - as the various findings discovered by the human "I" or human "awareness" suggest, for example, the age of the universe, fossil remains of other living beings etc., then they both, namely the universe and its source or the author called god or whatever, will remain inaccessible or unknown forever as far as human beings are concerned.

The nature of god and the nature of human 'I' are both inter-linked because they both are awareness'. The most fundamental difference that obtains between the two is that whereas god is an awe-inspiring disembodied awareness, human 'I', on the other hand, is a common-'o'-garden embodied awareness.

Cosmic Space is God and Physical Universe is God's Dream

~*~*~*~*~

THE ENTITY IN MAN WHO ADDRESSES ITSELF AS 'I' -2

The question, namely :-

'Who, or what is the source or the author of the existent universe besets mankind constantly

Before embarking on answering the above question, let's explore first who is the source or the author of the above enquiry in a human being?

This source or the author of the above enquiry is certainly not the physical body of man because physical body is a construct or the product or the produce or the merchandise of mere physical matter which has no innate or inherent ability to ask question because like all physical matter it is one hundred percent insensate or insentient.

Beside ghikphysical body, the only other entity comprising man is his 'I' or his awareness. This 'I' or awareness of man is well and truly or, if is preferred, naturally or inherently a sentient or sensate being, existent in man. As the 'I' or the awareness of man is well and truly or naturally or inherently a sentient or sensate being residing in man's physical body, it undoubtedly is the source or the author or the creator of the above enquiry.

Hence, man must always remain mindful of the fact that the questioner in him is always his sentient or senate 'I' and not his insentient or insensate physical body.

Likewise, man must always remain alert to the fact that the responder in man to any query or question, also is and always will be his 'I' and his 'I' alone and not his insentient or insensate physical body.

To sum up.

 Man must always remain vigilant to the fact that it is his 'I' and his 'I' alone and not his insentient or insensate physical body, who always is and always will be, both the questioner as well as the responder to any query or question.

SUPREMACY OF AWARENESS IN THE COSMOS -1

In addition to possessing an extremely obvious or easily perceived or understood or self-evident physical body, every human being also possesses an extremely subtle and mystical "awareness" plus an extremely subtle and mystical "conscience" associated with this extremely subtle and mystical "awareness".

In most cases, man is aware of his extremely subtle and mystical "conscience", even though he may or may not be actively cognizant of his extremely subtle and mystical "awareness". But the point to take note of here for man is that his extremely subtle and mystical "awareness" is the font of his extremely subtle and mystical "conscience".

Of course, it is true that some people are able to muffle or suppress and sometimes even extinguish their extremely subtle and mystical "conscience" by habitually acting against its advice. But even in such an undesirable scenario, human "awareness" continues to make its presence felt inside man's body by keeping the body sentient or sensate or illumined. Human "awareness" is the sap which keeps man's body sentient or sensate or illumined.

As said earlier, one's awareness is the author or the source of one's "conscience" or the "inner voice". As long as awareness continues to exist in man's body, so long as man's "conscience" continues to live on or abide in him even if he has succeeded in subduing or suppressing it entirely.

Man's awareness is not only the font of his "conscience" as well as the font of the sentience-giving sap for his body but it also is the true doer or the executor of all his volitional or willed tasks or activities.

All the volitional or willed tasks or activities which man undertakes   while alive, are performed by man's awareness only and no one else through   the agency or the medium of man's physical body. In other words, the true doer of man's all volitional tasks or activities is not man's intrinsically insentient physical body but his intrinsically sentient or sensate awareness and awareness only.

Furthermore, man's awareness is the spring-head of man's all thoughts, ideas, desires, dreams, ambitions and emotions.

It is amazing that even though the awareness of man is the source of supply of all his thoughts, ideas, feelings, dreams, ambitions and emotions plus volitional or willed deeds, or actions, how few human beings actually pay heed to their awareness. From time immemorial, this most puzzling behavior of man has been the greatest of all the riddles of this physical-matter-besotted or physical-matter-enamored or physical-matter-enthralled or physical-matter-captivated world.

If man ponders deeply and sincerely for a while upon his own awareness or, more correctly , if man's awareness for a while concentrates or focuses the entire might of its unique and incredible attention upon itself and nothing else , then , to its own great surprise , it will very quickly come to realize that 'it' 'itself'' is the very basis or the beginning or the fountain-head or the starting point, not only of man's all thoughts, ideas, desires, dreams, ambitions, emotions and actions, but in fact, it also is the very basis or the beginning or the fountain-head or the starting point of the entire conscious-existence of the cosmos as far as that particular human being is concerned.

It is extremely difficult for man's awareness to fathom or figure out the nitty-gritty or the most fundamental point of the self-evident truth of its own very existence or presence within man's body without the expense of a great deal time and effort on its part.

Isn't it extraordinary that this kind of baffling or confounding situation exists or operates not very uncommonly with regards to man's intrinsically self-illumined awareness?

SUPREMACY OF AWARENESS IN THE COSMOS - 2

Awareness is the most confounding entity there is anywhere in the cosmos. It is more mysterious than the most recondite marvel man will ever encounter anywhere in the cosmos.

No doubt, there is a surfeit of exciting and extraordinary beings, things, events and phenomena all over in the cosmos, but none of them will ever surpass or even match the magic and the mystery or the enchantment and the puzzlement of awareness.

Awareness is such a conundrum that if it can be unraveled to its deepest core, then the ultimate truth behind each and every being, thing, event and phenomenon of the cosmos will become patent to mankind. In other words, if man can successfully crack the secret of his own awareness, then the magic and the mystery of the incredible, timeless, dimensionless or formless god or the creator of the cosmos will also become patent to man.

Before the dawn of the modern scientific era, humanity did not have access to present day scientific gadgets. Hence, man then did not have any means to explore those physical riddles which were beyond the reach of his five physical sense organs. However, since the availability of present day, state-of-the-art scientific tools, lots of physical enigmas of nature, which were earlier hidden from man have now become accessible or intelligible or manifest to man.

Nevertheless, the fact remains that none of the scientific appliances invented by man till to date, irrespective of the height of their sophistication and precision, will ever be able to probe and solve the magic and mystery of the immaterial or insubstantial or non-physical awareness.

Man's five physical sense organs as well as his various scientific gadgets on which he is so dependent now a days for the purpose of exploring his physical milieu, are extremely crude as tools for deciphering the secret of the immaterial or the insubstantial or the non-physical awareness. Therefore, he exhibits extreme skepticism plus reluctance when he is told to explore such an immaterial or insubstantial or non-physical truth or reality as his own awareness which is beyond the ken of his physical sense organs as well as beyond the ken of his most advanced scientific gadgets.

On account of its immateriality or insubstantiality or non-physicality or on account of its non-dimensional nature or dimensionless-ness or formlessness, awareness is the most occult and the most profound truth there is anywhere in the cosmos which is yet to be understood in an entirely meaningful and fundamental way by humanity at large.

In most ways, man's awareness is like god's awareness. God's awareness as well as man's awareness are both self-aware or self-sentient or self-illumined truths meaning thereby they both are aware of their own being or of their truth or of their existence as an aware being or sentient being or sensate being or illumined being. Additionally, they both are aware of their own inner world of thoughts, ideas, desires, dreams and emotions which are very private or personal to them both.

However, man must always bear in mind that even though his awareness is like god's awareness in most of the ways, particularly with regards to its intrinsic nature or, if it is preferred, particularly with regards to its intrinsic qualities, however, the fact remains that, it merely is an extremely tiny part or an extremely tiny portion or an extremely tiny slice or an extremely tiny segment of god's awareness just as his physical body also is.

However, man's awareness must take cognizance of the fact that, being innately self-sentient or self-sensate or self-illumined or self-aware, it also is mankind's the one and only gateway to the incredible or awe-inspiring awareness of god, which is not the case with his innately insentient or insensate or un-illumined or unaware physical body. It is by exploring his own awareness first, that man can then subsequently negotiate successfully his way to god's awareness. This he will not be able to accomplish by exploring his physical body alone as he is doing at present on a very massive scale most of the time.

~*~*~*~*~

SUPREMACY OF AWARENESS IN THE COSMOS -3

Man's cold shouldering of his awareness, coupled with the habit of giving his best attention to the body and body's physical needs, is the greatest of all the puzzlements.

To some extent, this behaviour of man is due to the faulty training which his awareness metes out to itself as well as it receives from others around it, from the very early stages of its existence.

The very instant a human baby is born in the world, the whole attention of the family is directed towards this baby's physical form and its various traits such as gender, look, color of its skin, hair and eyes, its birth weight etc. This is followed immediately by their total preoccupation with that physical form's material needs and comforts, such as provision of suitable clothing to hide away its nakedness and to shield it from the inclement weather, catering of suitable food for its nourishment, if mother's milk is not available in adequate quantity , furnishing of an appropriate and safe crib for its repose, maintenance of a comfortable ambient temperature to its liking, and arrangement for the handling of the bodily wastes generated by it.

The awareness' of baby's kinsmen were themselves groomed and conditioned in exactly the same way by their parents and grand- parents from the time of their appearance in this world.

In this way, the perpetuation of the "Original Sin" called the "Sin Of Body-Consciousness" at the expense of "Awarenessbal-Consciousness" has continued from the time of dawn of man on this earth.

The terms "Awarenessbal-Consciousness" and "Body-Consciousness" have been employed here in order to pointedly contrast them against each other. This has been done with a view to bring home to man's awareness the matter of undeserved importance which it accords at present to the physical matter of the cosmos i.e. to all things physical in the cosmos including the physical body of man. The hope is that one day man's awareness will learn to accord to itself the importance which is long overdue as far as its own affairs are concerned, nay, as far as the affairs of the whole cosmos are concerned. Further hope is that it will stop according the exaggerated importance to the physical matter of the cosmos and to all the physical matter composed entities of the cosmos which is its unfortunate wont or habit or custom at present.

The lack of "Awarenessbal-Consciousness" on the part of man's awareness is firstly due to the lack of adequate crystallization on its part of the truth of its own existence inside man's insentient physical body as latter's motivating force. Secondly, it is on account of the lack of insight on its part of its own overwhelming importance not only in man's personal affairs but also in the overall affairs of the cosmos. Finally, it also is due to the overwhelming preoccupation of man's awareness with man's physical body and this body's physical needs. This is all very unfortunate indeed for man's awareness.

However, if human awareness, deliberately begins to nurture inside itself the absolute truth of "Awarenessbal-Consciousness", at the expense of "Body-Consciousness", through the vehicle of "self-training", then unbridled "Body-Consciousness" will begin to take a back seat or better still, will begin to gradually melt away and ultimately will take its leave from the recesses of man's awareness forever.

In such a scenario, man's awareness in question will naturally become fully trained to become mindful of its own constant presence inside the insentient physical body of man. As a consequence, its usual pre-occupation with the body and body's physical needs will reduce to a bare minimum.

The above can be put in another way.

The "Awarenessbal-Consciousness" as compared to "Body-Consciousness" is a state of human awareness wherein the latter has deliberately managed to fully train itself in a way that it has become fully cognizant of its own 'is-ness' or 'presence' or 'existence' inside the insentient physical body of man at the cost of its familiar and overwhelming enchantment with man's physical body and this body's physical needs.

The maintenance and the intensification of the "Awarenessbal-Consciousness" as opposed to the maintenance and the intensification of "Body-Consciousness" is a "sine qua non" or a necessary precondition for all those human beings who are in search or look-out for the source or the author or the creator of the cosmos called god or whatever.

If an individual, by intensive self-effort, begins to find itself   established or rooted or entrenched increasingly into the state of "Awarenessbal-Consciouness", then the awareness of such an individual, starts to take notice of its own "is-ness" or presence or existence more and more forcefully inside that individual's physical body. This leads to an inevitable reduction and finally total disappearance of the bane or the blight of "Body-Consciousness" from the recesses of the awareness of that particular individual.

The ambrosia or the nectar or the immortal food called the "Awarenessbal-Consciousness" must first be imbibed or quaffed and then thoroughly digested and internalized by man's awareness, if it is to embark upon the path of finding out who or what is the source or the author or the creator of the cosmos.

To recap.

The essence or the quintessence of the expression called "Awarenessbal-Consciousness" is that the awareness of an individual human being focuses its entire attention upon itself and not upon the physical body of the individual, the physical body inside which the awareness in question merely resides and reposes, thus, discovering along the way, during this process or journey, its own innate nature plus its own origin and function as well as its own extraordinary subtlety and complexity plus mystery and magic, untainted, even in the slightest, by the affliction or the scourge or the bane or the blight of "Body Consciousness".

The comprehension of the inner meaning of the term "Awarenessbal-Consciousness" on the part of human awareness will begin an extremely long and arduous, but nevertheless, an extremely exciting and fulfilling journey for the human awareness in question. This journey will take the latter into a realm, never entered by it before or accessed by it before. In fact, this voyage will transport human awareness into the "Supra-Physical plane" or "Awarenessbal Plane" or "Transcendental Plane" and ultimately even into the "Supra-Awarenessbal Plane" or "Supra-Transcendental Plane" or into the "Plane of the Author or the Creator of the Cosmos".

The "Supra-Physical Plane" or the "Awarenessbal Plane" or the "Transcendental Plane" is a level of thought cum existence of extraordinary clarity or lucidity for human awareness which always should or rather always must belong to human awareness because it is its innate or inborn inheritance or right but which it lost or mislaid due to its affliction or better still, due to its delusion of "Body-Consciousness".

To rcstatc.

The "Awarenesslism" or "Transcendentalism" or "Supra-Physicalism" is the very nature of human awareness, a nature which it lost or mislaid but lost or mislaid in its delusion and confusion only, but not at all in truth or reality or actuality.

Human awareness lost or mislaid its "Awarenesslism" or "Transcendentalism" or "Supra-Physicalism" on account of its overwhelming preoccupation with the physical body of man and total disregard or disdain of itself or better still, on account of its absolutely erroneous feeling that it itself is totally unworthy or undeserving of respect or consideration as compared to the mighty physical matter of the cosmos.

As mentioned at the very beginning of this chapter, man's birth from his mother's womb and the usual rituals associated with it, starts off a chain reaction, which leads to the increasing intensification of "Body-Consciousness" in human awareness at the cost of "Awarenessbal-Consciousness".

The subsequent growth & progress of human beings in the society and in the world at large continues to emphasize the same bias or partiality in favor of "Body-Consciousness"   at the expense of "Awarenessnessbal-Consciousness".

Thus, this unhealthy tradition or custom of 'Body-Consciousness', at the expense of "Awarenessbal-Consciousness" continues to thrive and and prosper, without any hindrance or hurdle or challenge or check from any quarter or source or person or place.

This is a great tragedy for human awareness as far as

its search of the absolute truth is concerned.

As said earlier, the scrounge of "Body-Consciousness", unwittingly causes the human awareness to overlook its own existence almost completely in the affairs of the cosmos. It also unwittingly causes the human awareness to turn almost a complete blind eye to its own pivotal role in the worldly discussions and discourses which it constantly undertakes on a massive scale, by necessity or perforce, during its sojourn into this cosmos in the company of the human body.

The above mentioned, extremely unhealthy dominance of the bane or the blight or the scourge or the affliction called "Body-Consciousness" in the affairs of the present-day cosmos has become so overpowering that it seems virtually impossible to get rid of.

But "get rid of it " is a must by the human awareness. And, the latter must replace it with the sublime truth of "Awarenessbal-Consciousness", if human awareness desires to move forward towards the goal of its unification with its source or the author or creator called god or whatever.

This goal is almost unachievable by the human awareness at the present moment on account of the rampant or the unbridled "Body Consciousness.

~*~*~*~*~

Dr. Chandra Bhan Gupta

SUPREMACY OF AWARENESS IN THE COSMOS - 4

At least one awareness of the intelligence-level of man or higher, is required in the physical universe in order to affirm independently the truth of existence of the physical universe.

Without the independent affirmation of the truth of existence of the physical universe by at least one human awareness, the reality of existence of such an insentient or insensate or mineral or material thing as physical universe will remain unconfirmed for -ever at least from the perspective of or at least from the point of view of human awareness - even if human awareness is prepared to grant that such a mineral universe or if one prefers, such a physical universe can exist in the absolute absence of human awareness.

Such is the critical importance of the human awareness as far as independent confirmation is concerned of the existence of the insentient mineral universe or the insentient material universe.

Despite the above described pivotal significance of the human awareness in regards to the confirmation of the truth of existence of the mineral or the physical universe, human awareness rarely gives any importance or weightage to itself in comparison to the mineral or the physical matter of the universe.

The bulk of the time of the human awareness is spent in thinking or ruminating plus reading, writing and talking about the mineral or the physical matter of the universe plus about all the items composed of this mineral or the physical matter of the universe.

Such is the overwhelming fascination of human awareness with regards to the mineral or the physical matter of the universe and all things mineral or physical that it almost completely forgets its own existence in this mineral or physical universe.

Not only this, human awareness overlooks completely its own pivotal significance or importance in this mineral or physical universe as far as the confirmation of the truth of existence of this mineral or physical universe is concerned.

What an amazing situation all this is?

When a human awareness is asked to describe a human being, it usually starts by portraying some of the characteristics of that human being's physical persona such as that persona's approximate age, height, girth, facial looks and any distinguishing mark it may have. Additionally, it may also describe that physical persona's educational qualifications and profession etc., if applicable. However, while doing so, at no point it bothers to take heed of the awareness present inside that physical persona and the fundamental importance of the existence of that awareness inside that physical persona, the awareness which made it possible for that physical persona to acquire such special tags as its educational qualifications and profession and the like.

Then there is a further point which also deserves attention on the part of human awareness and it is this:-

A human body which has lost its awareness permanently is called a dead body. It goes without saying that such a body becomes incapable of performing any conscious or willed or voluntary task which it was able to undertake while it was still in possession of its awareness.

Thus, its easy to appreciate the paramountcy of awareness with respect to all the conscious or willed or voluntary activities of the physical body.

And yet the deliberate thought with regards to the existence of human awareness inside the human physical or mineral body hardly ever occurs to human awareness in the manner it is clued into or is wise to regarding the insentient human physical or mineral body. This is very unfortunate indeed from the transcendental dimension or aspect, that is to say, from the dimension or aspect of human awareness. This situation is allowed to persist widely amongst mankind despite the fact that human awareness is the sole basis of all the conscious or willed or voluntary activities of the insentient human physical or mineral body.

Use of such expressions or labels as mind, intellect and ego etc. with respect to human awareness camouflages the latter still further and makes it more distant and mysterious than it already is. There cannot be any mind, intellect and ego without the awareness. Such terms merely refer to various functions of human awareness, nothing more nothing less.

Hence, human mind, intellect and ego must not be viewed as being different from human awareness. Human awareness encompasses them all.

Man, ordinarily is not cognizant or clued into or wise to the role played by his awareness in the conscious tasks undertaken by his insentient physical or mineral body. Sadly, this is the stark reality, despite the fact that the presence or the existence of human awareness inside the physical or the mineral persona of man is the 'sine qua non' or the indispensable pre-condition for man's conscious existence. Conscious existence of man differentiates him from such an inherently insentient existence as rock or mineral.

During his conscious existence, man is aware of his own existence as an aware being as well as the existence of all the other beings, both of the aware kind as well as of the insentient or insensate kind or of the mineral or the material kind.

It is during this conscious existence that man has the chance, if he so wishes, to become curious about the ultimate source of the mineral or the material universe as well as the source of its own existence.

Whether man realises or not, the truth is that he has only one tool in his possession to approach the source of the mineral or the material universe as well as the source of the awarenessbal or the consciousnessbal universe and this tool is his awareness and awareness only. There is nothing else to aid him in this regard. Hence, he must use the services of his awareness to the full if he wants to achieve the real goal of his life which is to discover who is his ultimate source or the author or the creator.

From what has been said above, it should now be possible for human awareness to appreciate its special significance over and above that of his insensate physical or mineral body inside which it merely resides and reposes, nothing more nothing less.

It is very unfortunate for human awareness that it spends most of its precious time in giving precedence to its physical or mineral body over itself. The undesirable consequence of this behaviour on its part is that it hardly ever wins the full focus of its own care and attention.

Since the existence of human awareness inside a human physical body is a "sine qua non" or an indispensable pre-condition for all the willed or volitional activities & attainments of that particular human physical body, human awareness therefore must logically be the chief target of its own constant care and attention plus exploration and investigation. Sadly however, this is not the case.

As said before, human awareness, instead spends most of its valuable time in looking after the physical or the mineral body and its needs or demands and also in exploring and investigating the physical or the mineral matter of the universe and all things physical or mineral, all of which are inherently insentient or insensate in nature without exception. What a sad state of affairs?

It does not behove human awareness to behave in such an unthinking manner, human awareness which, unlike physical or mineral matter is a sensate or sentient or cognisant or knowledgeable or tuned in being and not only this, which additionally also is the most intelligent being on this earth.

WHAT GOD TRULY IS - 1

It is but natural for one to ask :- "What god truly is"

Hence, this question must be answered.

Expectation is that out of this answer, will emerge enough information to put to rest all one's doubts and uncertainties regarding god. This answer should also include intelligible or uncomplicated information for one with regards to the method employed by god to bring forth the formation of the present-day 3-D or the three-dimensional universe out of its own substance.

Mankind possesses a staggering amount of information with regards to the physical or the mineral manifestation of god i.e. the present-day 3-D or three-dimensional physical or mineral universe. In contrast, however, it possesses very little information as to what god truly is when it is unmanifest or what god truly is prior to giving rise to the 3-D or the three-dimensional physical or mineral universe or better still, what god truly is in its original or primal or primeval or pristine state.

Therefore, one's attempt must be to reach out to the true essence of god when the latter is totally unmanifest or when it is in its original or primal or primeval or pristine state.

First of all, let one convey or make known that god is an awareness and nothing but an awareness. However, it is quite unlike human awareness. Human awareness is an embodied awareness whereas god's awareness, in stark contrast, is an incredible disembodied awareness.

God's incredible disembodied awareness or, if one prefers, god's incredible bodiless awareness is capable of existing in two forms.

But one has to bear in mind one extremely important fact and it is this :-

That god's incredible disembodied or bodiless awareness never exists in both forms simultaneously or concomitantly or concurrently or at one and the same time. Instead, at any given instant, it exists in either in one form or the other.

These two forms of god's incredible disembodied or bodiless awareness are the following :-

Dimensionless form.

3-D or three-dimensional form.

1. Dimensionless Form of God's Incredible Disembodied Or Bodiless Awareness

In its original or primal or primeval or pristine or fundamental state, god's incredible, disembodied or formless awareness exists as a "dimensionless awareness".

Since in its original or primal or primeval or fundamental state, god's incredible, disembodied or formless awareness exists as a "dimensionless awareness", it does not manifest or display or exhibit, while in this state, inside itself any kind of 3-D or three-dimensional cosmic space nor any kind of 3-D or three-dimensional physical or mineral universe in the manner it is busy doing at present.

What humanity today labels as physical or mineral or material universe, it is in fact the collective name of all the 3-D or the three-dimensional forms which are being manifested or displayed or exhibited by god inside its own incredible disembodied or formless awareness and they all have been generated or produced by god inside its own incredible, disembodied or formless awareness through the instrumentality or the activity of daydreaming on its part.

Since god's incredible, disembodied or formless awareness, in its unmanifest or original or primal state, is a "dimensionless awareness", physically or minerally embodied human awareness finds almost impossible to picture or imagine such an extraordinary, unmanifested state of god's incredible disembodied or formless awareness wherein the latter abides in a wholly "dimensionless state".

God's incredible, original or primal persona or identity of pure or pristine disembodied or formless awareness wherein it abides in an absolutely "dimensionless state" is extremely difficult to imagine or picture or visualise or conceptualise by the embodied human awareness.

However, embodied human awareness must try hard to grasp this difficult to imagine or picture, "dimensionless primal persona or identity" of god's disembodied or formless awareness if they wish to advance beyond their unexciting or mundane, mineral or material mask or the exterior called the physical body.

2. 3-D or Three-Dimensional Form Of God's Incredible Disembodied or Formless Awareness

Above one has described the nature of god's incredible, disembodied or formless awareness in its original or primal or primeval or fundamental state which, as said previously is that of a "dimensionless awareness".

From time to time, this original "dimensionless and disembodied" awareness of god decides to daydream.

Before this "dimensionless and disembodied" awareness of god can embark upon its activity of daydreaming, it perforce or by necessity, has to first expand or extend or dilate or distend or inflate or enlarge its original "dimensionless awareness". This it has to do in order to form or create inside its awareness sufficient amount or quantity of a 3-D or three-dimensional, "awarenessbal-space" or "awarenessbal-acreage" or "awarenessbal-territory" so as to spatially or territorially accommodate inside itself, its subsequently created 3-D or three-dimensional, daydream-stuff composed universe which is called or labelled as physical or mineral or material universe by human awareness on account of its total ignorance or nescience of the true state of affairs in this regard.

Thus, what human awareness calls or labels as cosmic space is, in reality or truth, the awe-inspiring, "awarenessbal-space" or "awarenessbal-acreage" or "awarenessbal-territory", existing or abiding inside the incredible, disembodied or formless, awareness of god, created or formed by the latter by expansion or distension or dilation or inflation of its awareness, nothing more nothing less.

Furthermore, what is called or labelled as physical or mineral or material universe by human awareness, truly is a daydream of god's incredible, disembodied or bodiless awareness, nothing more nothing less.

In other words, what is believed and accepted by human awareness as hard, solid and real plus physical or mineral or material, is no more hard, solid and real plus physical or mineral or material than any daydream.

WHAT GOD TRULY IS - 2

In Adwait-Vedanta, god or the source or the author or the creator of space time and physical universe is described in Sanskrit language as having three core characteristics, namely:-

Sat

Chit &

Ananda

   Or

Existence

Awareness &

Bliss or Joy. Here the word bliss or joy is the stand-in for surrogate for or substitute for or proxy for or a representative case for the word emotion or feeling or sentiment. Emotion or feeling or sentiment is an intrinsic essence or quintessence of all the aware beings or of all the awareness', irrespective of whether the awareness in question is the incredible, disembodied or bodiless awareness of god or the common-'o'-garden embodied or in the flesh awareness of man.

People who are engaged in the calling or the mission of exploring   god or the source or the author or the creator of space time and physical universe are likely to be familiar with the above described Vedantic triad of *Sat, Chit and Ananda* or *Satchitananda.*

WHAT GOD TRULY IS - 3

From the Adwaitic perspective, a human being is made up of two key or cardinal components namely the physical body and the non-physical awareness.

The physical or the material human body is transient or temporary or time bound. In contrast, the non-physical or non-material, human awareness is amaranthine or eternal or immortal.

The transience or impermanence of the human physical or material body contrasts sharply with the immortality or deathlessness or timelessness of the human non-physical or non-material awareness.

Human awareness is immortal in the manner of the awareness of god or in the manner of the awareness of the source or the author or the creator of the physical universe. This is so because human awareness is a "pristine" form of a segment of god's immortal awareness.

Being a "pristine" form of a segment of god's immortal awareness, makes the human awareness also, immortal in the manner of god's awareness because the two are identical or one and the same in terms of most or majority of their intrinsic qualities including the quality of immortality or timelessness.

On the other hand, however, human physical body as such or, human physical body as merely a human physical body, is as mortal as it seems. This is so because, in contrast to the immortal human awareness, which is a "pristine" form of a segment of god's immortal awareness, human physical body is a "condensed" form of a segment of god's immortal awareness.

Being a "condensed" form of a segment of god's immortal awareness makes the human physical body as mortal as it seems.

However, one must take serious notice of one fact here and it is this:-

The word "mortal" which has been employed above in the context of the human physical body, has been made use of, not in an "absolute sense", but merely in a "relative sense".

This is so because in the "absolute sense" or at the "supreme level", immortal human awareness as well as mortal human physical body are both nothing but the immortal awareness of god only, even if one amongst them, namely the immortal human awareness is a "pristine" form of a segment of god's immortal awareness whereas the other, namely the mortal human physical body is a "condensed" form of a segment of god's immortal awareness.

It is the process of "condensation" or if one prefers, it is process of "compaction" or "compression" of a segment of god's immortal awareness, which has resulted in the conversion of the latter into a human physical body.

However, this very same process i.e. the process of "condensation" or "compaction" or "compression" of a segment of god's immortal awareness which has caused the conversion of the latter into a human physical body, has also made the human physical body "relatively" or "comparatively" mortal, that is to say, relative to or compared to the immortal human awareness which, as said before, is a "pristine" form of a segment of god's immortal awareness.

To sum up.

Human physical body as such or human physical body merely as human physical body is mortal.

However, at the level of the supreme essence or quintessence, that is to say, at the level of the immortal awareness of god, which has composed or constituted this human physical body out of its own very self, this human physical body is not strictly mortal. Instead, one must appreciate here clearly that it is as immortal as the supreme essence or quintessence in question namely the immortal awareness of god which has composed or constituted this human physical body out of its own very self or out of its own very awareness.

The above can be put in another way.

At the supreme level, an individual human physical body is also a segment of god's immortal awareness and nothing but a segment of god's immortal awareness, notwithstanding this segment's condensation or compaction or compression in order to give rise to or in order to construct or build a human physical body.

In other words, since at the supreme level, human physical body too, is a segment of god's immortal awareness and nothing but a segment of god's immortal awareness, albeit a "condensed" segment of god's immortal awareness, it is immortal and immortal only and nothing but immortal and immortal only.

To repeat.

An individual human physical body as merely an individual human physical body is mortal. No one can deny this. But at the highest or the supreme level, it is a segment of god's immortal awareness albeit a "condensed" segment of god's immortal awareness. Hence, from this perspective, that is to say, from the fundamental perspective of the essence or the quintessence which has composed this human physical body, the latter is as immortal as this essence or quintessence i.e. god's awareness.

If the immortal human awareness wants to acquire the knowledge regarding the immortal awareness of god, that is to say, regarding its own immortal source as well as the immortal source of the transient human physical body, inside which it is temporarily residing and reposing at present, then the exploration and the analysis of the transient physical universe by it will be of no help.

Immortal human awareness instead, has to look towards the immortal awareness of god only or towards the immortal awareness of its source only if it wants to gain knowledge about the latter which, as said earlier, is the source or the author not only of its own immortal self but also the source or the author of the entire, seemingly mortal or the seemingly transient physical cosmos.

To sum up.

Immortal human awareness should not therefore, spend its precious time in exploring and analysing the transient or the ephemeral physical or the material universe if its goal is to unravel or uncover the secret of the immortal awareness of god or the secret of the immortal awareness of the source of the transient or the ephemeral physical or material universe as is its wont or the habit or the custom at present.

Presently the scientists are intensely involved in the search of the answer with regards to the origin of the present-day, ephemeral physical universe with the help of their such ephemeral but extraordinary physical gadgets as the large hadron collider and the like.

Only the immortal, non-physical awareness of god or the immortal, non-physical awareness of the source of the ephemeral physical universe is in a position to oblige mankind with the enlightenment it is seeking with regards to the origin of the present-day, ephemeral physical universe and not their ephemeral but extraordinary physical gadgets such as the large hadron collider and the like.

The supra-transcendental knowledge, that is to say, the supra-awarenessbal knowledge about the immortal awareness of god is impossible to be had or is impossible to get one's hand on, through the study and analysis plus experimentation on the ephemeral physical matter of the cosmos as is the wont or the habit or the custom or the practice of the present-day scientists.

The study and analysis plus the experimentation on the ephemeral physical manifestations of the immortal non-physical awareness of god will take human awareness nowhere near the immortal non-physical awareness of god. Instead, they all will push human awareness more and more away from the immortal non-physical awareness of god or, better still, they all will create more and more distance or gulf or schism or rupture between the human awareness on one hand and god's awareness on the other.

In other words, such explorations by the human awareness of the ephemeral physical manifestations of god's immortal non-physical awareness, will take the human awareness, more and more close to these ephemeral physical manifestations of the immortal non-physical awareness of god, and not the other way around, that is to say, and not towards the immortal and non-physical awareness of god whom the human awareness seeks and from whom all the ephemeral physical manifestations of the cosmos have emerged.

WHAT GOD TRULY IS - 4

God's immortal awareness can be appraised and describedby the immortal human awareness from two contrasting perspectives.

These are:-

Appraisal and description of god's immortal awareness by the immortal human awareness from the position that god's immortal awareness is carrying or sustaining or, better still, exhibiting or displaying inside its immortal awareness its transient awarenessbal daydream or its transient awarenessbal reverie or dreamry or imagery or phantasy called the physical universe. This is the factual state of affairs or factual situation with regards to the immortal awareness of god at this instant in time.

Appraisal and description of god's immortal awareness by the immortal human awareness from the perspective that god's immortal awareness is in its original or primal or primeval or pristine state i.e. that god's immortal awareness is not carrying or sustaining or exhibiting or displaying inside its immortal awareness its transient awarenessbal daydream or its transient awarenessbal reverie or dreamry or imagery or phantasy called the physical universe. This was the factual situation or the factual state of affairs with regards to the immortal awareness of god, 13.7 billion light years ago for this is the total age of the present-day physical universe at this instant in time, as per the evidence gathered and analysed by the immortal human awareness of today.

Let's deal with the above two positions vis-a-vis god's immortal awareness, one by one.

Appraisal and description of god's immortal awareness by the immortal human awareness from the position that god's immortal awareness is carrying or sustaining or exhibiting or displaying inside its immortal awareness its transient awarenessbal daydream or its transient awarenessbal reverie or dreamry or imagery or phantasy called the physical universe.

Appraisal and description

The present-day physical universe is an awarenessbal daydream or is an awarenessbal reverie or dreamry or imagery or phantasy of god's immortal awareness, nothing more nothing less. This is so despite the fact that physical universe is god's extraordinary awarenessbal product or awarenessbal construct of breathtaking depth and design plus detail and definition as well as variety and diversity which will never be matched by any human awareness. Human awareness cannot match or equal god's awareness in this regard because of latter's size, might, power and intelligence or, better still, because of latter's size, might, power plus imagination and intelligence.

What has been said above can be expressed in another way.

Ephemeral physical universe has been granted its transient existence by god's immortal awareness inside its own very self, that is to say, inside its own very immortal awareness and nowhere else. This is so, because the truth is that the ephemeral physical universe has been created by god's immortal awareness through the process or the activity of daydreaming on its part and nothing else.

It will not be at all unwarranted if one repeats here that the ephemeral physical universe of today has been created by god's immortal awareness through the process or the activity of daydreaming on its part and nothing else.

To recap.

The ephemeral physical universe and all its contents, both animate and inanimate plus all its events or incidents or bashes or episodes, are mere awarenessbal phenomena or mere awarenessbal products or mere awarenessbal merchandise or mere awarenessbal creations, existing inside or within the immortal awareness of god only and nowhere else, which god's immortal awareness enjoys watching inside its own awareness and derives considerable pleasure by doing this. To put all this very simply. Physical universe has been created by god's immortal awareness for its own amusement or enjoyment. The entity labeled as hard, solid, physical universe by the immortal human awareness is merely an awarenessbal phenomenon or an awarenessbal product or an awarenessbal daydream or an awarenessbal reverie or dreamry or imagery or phantasy of god's immortal awareness which the latter has created inside its own awareness for its own entertainment and amusement, nothing more nothing less.

Hence, the ephemeral physical universe and all its contents and events, that is to say, all its beings, things and occurrences exist or abide or reside or have their spatial or territorial placement inside god's immortal awareness only and nowhere else. Furthermore, the entity called the "cosmic space" by the immortal human awareness, inside which the physical universe has its spatial-placement or territorial-placement, in truth or reality, is the "awarenessbal-space" or, if one prefers, in truth or reality, is the "mind-space" of god and nothing else. "Awareness" and "mind" are two names of the same thing or are one and the same thing and not two different things in the realm of Adwait-Vedanta.

What has been said above can be put in another way.

God's "mind-space" or god's "awarenessbal-space" is the entity called the "cosmic space" by human awareness and thus "cosmic space" has its existence or presence or "is-ness" inside god's immortal awareness or god's immortal mind only and nowhere else.

Appraisal of god's immortal awareness by the immortal human awareness from the perspective that god's immortal awareness is in its original or primal or primeval or pristine state i.e. when god's immortal awareness is not carrying or sustaining or exhibiting or displaying inside its immortal awareness its transient awarenessbal daydream or its transient awarenessbal reverie or dreamry or imagery or phantasy called the physical universe.

Appraisal and description

The highest knowledge for immortal human awareness consists of knowing what god's immortal awareness really is when it is not carrying or sustaining or exhibiting or displaying inside its immortal awareness its transient daydream or its transient reverie or dreamry or imagery or phantasy called the physical universe.

This knowledge must enable immortal human awareness to cut-through or pierce-through or force a way through all the layers of camouflage which god's immortal awareness has wrapped around itself in the shape or the guise of physical universe.

In other words, such an insight must enable the immortal human awareness to comprehend the immortal awareness of god in its totally un-camouflaged or undisguised or unconcealed form. As mentioned earlier, this camouflage or disguise or concealment around the "face" of god's immortal awareness is none other the well- known transient entity called the physical universe.

This camouflage or disguise or concealment namely the present-day physical universe has been wrapped by the immortal awareness of god around itself in order to play a kind of hide and seek game with the immortal human awareness, which is the most brilliant amongst all the awarenessbal-progenies or all the mind-progenies of the immortal awareness of god till to date or until now.

Thus, the camouflage or the disguise or the concealment of god's immortal awareness, called the transient physical universe, is an "awarenessbal-progeny" or a "mind-progeny" of the immortal awareness of god and is therefore, made up of or composed of god's awarenessbal-stuff" or "mind-stuff " only and nothing less. That is to say, it is not at all made up of or composed of what immortal human awareness ignorantly calls hard and solid physical matter.

The stuff of which the entity called physical universe is made up of or composed of is no more hard, solid and physical than any daydream-stuff.

Now let's deal with the issue of the nature of god's immortal awareness   13.7 billion light years ago, that is to say, before it formed or created or gave rise to inside its own immortal awareness, it's amazing awarenessbal daydream or awarenessbal reverie or dreamry or imagery or fantasy called the physical universe through the process or the activity of daydreaming on its part.13.7 billion light years ago, that is to say, before it formed or created or gave rise to its awe inspiring or mind blowing awarenessbal daydream or awarenessbal reverie or dreamry or imagery or fantasy called the physical universe inside its own immortal awareness, god's immortal awareness existed as an incredible, pure or pristine "timeless dimensionless awareness".

UNVARNISHED FACE OF GOD OF THE UNIVERSE

God or the creator of the physical universe, in its fundamental form is an incredible timeless dimensionless awareness or an incredible timeless dimensionless mind of limitless intelligence.

It is gender neutral and therefore is referred to as 'it'.

Before the beginning of time and after the end of time it exists in its incredible fundamental form described above i.e. as an incredible timeless dimensionless awareness or as an incredible timeless dimensionless mind of limitless intelligence.

However periodically after the beginning of time and

before the end of time it metamorphoses itself into a transient four- dimensional space time and an accompanying transient 3-D or three-dimensional physical universe.

One will tell later how it metamorphoses itself into a transient 4-D or four-dimensional space time and a transient 3-D or three-dimensional physical universe but at this juncture one will like to say that it has no progenitor or first cause and is singular or one and only.

During the course of its lone eternal existence as a timeless dimensionless awareness or as a timeless dimensionless mind, it sometimes , in order to love itself by itself , separates a part of itself within self from its whole-self and experiments and forms inside itself or if one prefers inside its timeless dimensionless awareness or timeless dimensionless mind a transient 4-D or four dimensional space time and an accompanying transient 3-D or three-dimensional physical universe as its two transient awarenessbal imageries or as its two transient mind or mental imageries which are labeled as cosmic space time and physical universe respectively by man .

These two transient awarenessbal imageries or mind or mental imageries of creator namely the 4-D or four-dimensional space time and accompanying 3-D or three-dimensional physical universe, formed by the creator and existing inside the creator or, if one likes, formed by the creator and existing inside the creator's incredible timeless dimensionless awareness or incredible timeless dimensionless mind , are real but real only in the sense all awarenessbal imageries are real or all mind or mental imageries are real .

They however are not real in the sense human awareness takes them to be or believes them to be.

To put this in another way.

Creator's two transient or impermanent creations in question namely the 4-D or the four-dimensional space time and accompanying 3-D or three-dimensional physical universe, both of which have been formed inside creator by creator and therefore exist inside creator and by the wish or will or fancy or mood of creator, are real but real only in the limited sense of two transient awarenessbal imageries of creator or in the limited sense of two transient mind or mental imageries of creator.

They however are not real in the sense of man's understanding of the word "real" which is that physical universe is composed of a hard and solid reality called matter.

The so called hard and solid reality of matter is of the same degree of reality as that of any awarenessbal imagery or any mind or mental imagery albeit awarenessbal imagery or mind or mental imagery of creator, creator who in its fundamental form is an incredible timeless dimensionless awareness or timeless dimensionless mind of infinite intelligence.

At this point one will like to draw one's attention to one very startling fact which one doubts very much one has been able to surmise from what one has told already so far and it is this :-

The entity which humanity labels as cosmic space is none other than this very awareness or mind of creator albeit awareness or mind of creator in its expanded or distended or extended or enlarged form or, better still, in its inflated form, inside which the entire mankind along with the rest of the 3-D or three-dimensional physical universe is floating or wafting or levitating plus whirling or twirling or spiralling as a mere transient or impermanent awarenessbal or mental imagery or daydream of creator .

In other words, cosmic space from now onwards should be called the awareness or the mind of creator or should be re-named as the awareness or the mind of creator or, better still, should be re-named as the "awarenessbal-space" or the "mind-space" of creator by all mankind and its older name "cosmic space" should be thrown into a trash bin.

Furthermore, all human beings should re-name themselves as "awarenessbal-children" or "mind-children" of creator or god.

What has been described above is the most amazing truth. Now mankind should have no problem in seeing and finding god in everyone in the cosmos and everywhere in the cosmos all the time.

One will like to close by saying that, after a time, creator will end the time itself as well as its two inseparable companions namely the 3-D or the three-dimensional cosmic space and the 3-D or the three-dimensional physical universe. Creator will achieve this feat i.e. ending of time, ending of cosmic space and ending of physical universe by absorbing them all into itself by simply stopping imagery making. After the absorption and thus ending of these three transient imageries, creator will re-assume its basic form namely that of a timeless dimensionless awareness or timeless dimensionless mind.

It will remain in this form till it once again decides to begin another round or cycle of formation of the imagery called the 4-D or the four-dimensional space time and the accompanying 3-D or three-dimensional physical universe inside itself.

Such repeated cycles of formation and absorption of the imagery called 4-D or four dimensional space time and associated 3-D or three-dimensional physical universe goes on ad infinitum inside the creator as per creator's mood or fancy.

This is all there is to the so- called creation and destruction of space time and physical cosmos, no more, no less.

CREATOR-CREATED DUO-AN ETERNAL CONTINUUM

Immortality is the hallmark or the most distinctive feature of existence or being or truth or reality. It eternally see-saws or oscillates between its two manifestations or expressions called god on one hand & the universe on the other or the "Creator" on one hand & the "Created" on the other.

Existence, being, truth, or reality is eternal meaning thereby there never is and there never will be the state called "nothingness" or the state called "cipher" or "zero" or "shoonya".

There are two phases of existence which are as follows :-

One phase of existence obtains without having space-time in its ambit. This phase is called god or the "Creator". The other phase is space-time inclusive and is called the "universe".

Universe

Let's discuss the term "universe".

The term "universe" encompasses not only the space-time but also the physical matter which is extant in the space-time plus the associated awareness extant in the space-time such as human awareness. The term space-time refers to the twosome consisting of the cosmic space and the cosmic time.

Space-time exclusive and space-time inclusive phases of existence

The space-time exclusive and the space-time inclusive phases of existence are not two different things. Universe is god or the "Creator" in metamorphosed or transmuted form. After a while universe reverts back to its native or original or primeval condition called god or the "Creator".

Metamorphosis or transmutation of existence

Metamorphosis or transmutation of existence is a common-'o'-garden or everyday, commonplace or run-of-the-mill process in the universe. Just look around. Thus, the "Creator-Created" duo or the pair is an eternal continuum or an eternal continuance or continuity or perpetuity or constancy or inseparableness or inextricableness, nothing more nothing less.

From what has been said above it transpires that "Creator-Created" duo or pair is merely a seeming or apparent or ostensible duality. It is not an "Absolute Duality" in the sense that "Twain shall never meet". At the absolute level or, if one prefers, at the supreme level, the "Creator-Created" duo or pair is one and the same thing or one and the same being or one and the same truth and not two things or two beings or two truths. Here one will like to re-emphasise that at the supreme level, the "Creator-Created" duo or pair is one and the same thing or one and the same being or one and the same truth and not two things or two beings or two truths. This is the core or the key or the fundamental or the central meaning of the term or the word "Adwaita".

Sadly, however, this duology or duality or duo or dyad called the 'Creator-Created' pair or the couple or the team or the yoke is not accepted and internalised as merely a seeming or an apparent or an ostensible duality or duology by mankind. Instead, it is perceived by mankind as well as genuinely accepted and internalised by mankind as an absolute duality or duology with the sense or the feeling that the "Twain shall never meet".

But this is not the case with the genuinely wise who are fully awake to the fact of **"Absolute Unity Amidst The cacophony Of Diversity"** in the universe.

Supreme truth is one only but the wise describe it in different ways.

This supreme truth of **"Absolute Unity Amidst The Cacophony Of Diversity"** must be self-discovered. It must not be faith-based or religion-based truth because faith-based or religion-based truth differs from one faith to another or from one religion to another.

Existence always exists but sometimes it exists as "Creator" and at the other times as "Created". The "Creator" is the "Created" in the metamorphosed or the transmuted form. The "Created", after its absorption or dissolution into the "Creator", becomes one with the "Creator". Or more to the point, the "Created", after its demise or disintegration or termination or cessation inside the awareness or the mind of the "Creator" becomes one with the awareness or the mind of the "Creator".

There never is and there never will the state of "nothingness" or the state of "cipher" or "zero" or "shoonya".

Paradoxically, the existence - whether it is at the level of "aware God" i.e. the "aware Creator", or at the level of the "universe" i.e. the "Created" - always starts from the state of "Absolute Unity" and then transmutes or metamorphoses itself into duality or diversity or multiplicity and finally, after a while, reverts back to the state of "Absolute Unity" from where it started in the first place.

This is the supreme truth no matter from what angle or from what height one looks at this supreme truth.

Creation is diversity in unity.

ABSOLUTE UNITY AMIDST THE CACOPHONY OF DIVERSITY

The physical universe is an amazing entity, both from the point of view of variety as well as the number of beings it contains, not to mention the countless interactions which take place between them at any given moment in time.

As a result, it is very difficult for some human awareness' to believe and accept that all this astonishing diversity they perceive and experience in the present-day physical universe has originated or emanated or has come out of one single source or being or author or creator. However, the supreme truth with regards to the origin of the present-day physical universe is exactly or precisely, **this very one,** which is that the present-day mammoth and marvellous or epic and extraordinary physical universe did originate or rather has originated from or emanated out of one single source or one single being or one single author or one single creator. Unfortunately, though, as said before, this supreme truth, many amongst the human awareness' find extremely hard to believe and accept.

The reason, these human awareness' find it extremely hard to believe and accept that the source or the author or the creator of the present-day physical universe is one, and only one, supreme being is that the scope or horizon of these human awareness' has been wilfully or deliberately constrained or curbed by this very same, one and only one, supreme being or creator in order that diversity is obtained in the present-day, polychromatic or kaleidoscopic physical universe with regards to the views or opinions about the origin of the physical universe.

This amazing one and only supreme being which has authored or created this amazing multi-cored or prismatic physical universe can be called by any name one fancies including the name god or whatever. The name by which this extraordinary supreme being is called or addressed is of no importance or significance at all. But, what is of the utmost importance or significance for the human awareness' to know is that this unique or one and only supreme being is an aware being or a sentient being & not an unaware or insentient being in the manner of a lump or a mass or a wad of physical matter called singularity or cosmic egg or whatever as opined by the present-day scientist of our planet earth.

The Supreme Being aka creator of the physical universe

The native nature or the innate nature or the original nature of the supreme being or the creator or the author of the physical universe is that of an incredible **"Timeless Dimensionless Awareness".**

The one and only entity which, by its very nature or as per its innate or essential quality or character, can be and is in fact **dimensionless,** is the **awareness** and nothing but the **awareness,** irrespective of whether the **awareness** in question is that of the source or the author or the creator of the present-day physical universe or that of a human being.

An awareness - irrespective of whether it is the awareness of the source or the author or the creator of the physical universe or that of a member of the human species - by being innately or connately **dimensionless,** does not need or require or better still, is not in the need or the requirement of a **pre-existent space** or territory or area or acreage for its own spatial or territorial placement and existence. And hence, it can be and in fact, is **the one and only entity** anywhere which can be and in fact is the creator or the progenitor or the begetter or the source or the author or the maker of space or territory or area or acreage of any size or scope or scale or capacity or immensity.

Therefore, the truth is that, the awe-inspiring and one and only **awareness** which is the source or the author or the creator of the present-day physical universe is also the source or the author or the creator of the present-day space or territory or the area or acreage called **cosmic space.**

In other words, the incredible or awe-inspiring **awareness,** which is the source or the author or the creator of the present-day physical universe, is not only the source or the author or the creator of the latter namely the physical universe but it also is the source or the author or the creator of the present-day, fantastic or breathtaking space or territory or area or the acreage called **cosmic space; cosmic space** inside which the present-day physical universe is not only housed or placed or accommodated but it also is floating or wafting or levitating plus whirling or twirling or spiralling non-stop from the very beginning of the present-day **cosmic time.**

By the way, the incredible or awe-inspiring **awareness,** which is the source or the author or the creator of the present-day **physical universe** as well as the source or the author or the creator of the present-day **cosmic space,** has created or constructed or forged or formulated the latter, namely the **cosmic space** inside itself by the process of **dilation or distension or expansion or extension or enlargement or inflation** of its own **original or primal or pristine, dimensionless awareness**. Thus, without any further ado or hubbub or brouhaha, let one inform the human awareness' that this **Supreme Being aka god or whatever** is not only the source or the author or the creator of the present-day physical universe but in fact it also is the source or the author or the creator of the present-day, fantastic or breathtaking space or territory or area or acreage called c**osmic space; cosmic space** which spatially or territorially houses or accommodates the present-day, **3-D or three-Dimensional physical universe** inside itself. In other words, the entity called **cosmic space** by human awareness' is the **dilated or distended or expanded or extended or enlarged or inflated** form or incarnation of the **original or native, Dimensionless Awareness** of **The Supreme Being** in question namely the source or the author or the creator of the present-day physical universe.

To repeat.

What is labelled as **3-D or the three-dimensional**

"Cosmic Space" by human awareness', in fact is the **dilated or distended or expanded or extended or enlarged or inflated** form or incarnation of the original **Dimensionless Awareness** of the **Supreme Being** in question namely the source or the author or the creator of the present-day physical universe.

Hence, **Cosmic Space,** in truth or reality is not some inert or insentient or insensate or inanimate or indolent or dull or unconcerned or senseless thing or entity which somehow or better still, which by serendipity or by fluke or by happy coincidence or by chance or by good luck or by good fortune happened to be there at the precise or at the exact moment in order to spatially or territorially house or accommodate the present-day **3-D or three-dimensional** physical universe.

In absolute contrast to what has been said above, the **cosmic space,** instead is an incredible or awe-inspiring, ubiquitous and the infinite **Field of God's Awareness** or the ubiquitous and the infinite **Field of Awareness of the Supreme Being** in question namely the source or the author or the creator of the physical universe.

Let's now discuss the **dimensional nature** or better still, **the 3-D or three-dimensional nature** of the physical matter of the cosmos.

The physical matter of the cosmos, in contrast, that is

to say, the physical matter of the cosmos in comparison to the awareness of the cosmos - irrespective of whether the awareness in question is that of the creator of the cosmos or that of a human being - always is and always will be a **dimensional entity** or better still, always is and always will be a **3-D or three-dimensional entity**, no matter how small or how little in size.

The implication of what has been said above with regards to the physical matter of the cosmos is that the latter, namely the physical matter, by being **dimensional** in nature or in its innate or essential quality or character, no matter how small or little in size, always is and always will be, in need of or requiring of, a pre-existent space or territory or area or acreage for its spatial or territorial placement and existence. In other words, physical matter cannot exist and will never exist without pre-presence or pre-existence of space or territory or area or acreage for its spatial or territorial placement and existence. That is to say, pre-existence or pre-presence of a **3-D or three-dimensional space** or territory or area or acreage is a sine qua non or an indispensable pre-condition for the presence or the existence of the **3-D or three-dimensional physical matter** in the cosmos. This will be the case even if the lump or the mass or the wad or the dollop of the physical matter in question is of an infinitesimally small size. If the lump or the mass or the wad or the dollop of the physical matter in question is of an infinitesimally small size, then even in this scenario, this infinitesimally small lump or mass or wad or dollop of the physical matter will still require a pre-present or pre-existent space or territory or area or acreage for its spatial or territorial placement and existence albeit a pre-present or pre-existent space or territory or area or acreage of only an infinitesimally small size.

From what has been said above, the conclusion with regards to the physical matter of the cosmos therefore, is that it will always need or require a pre-present or pre-existent space or territory or area or acreage for its spatial or territorial placement and existence, irrespective of its size or scope or scale or capacity or immensity. That is to say, irrespective of whether its size or scope is infinitely huge or mammoth or infinitesimally small or little. In other words, physical matter cannot exist in the cosmos in the absence of a pre-present or pre-existent space or better still, in the absence of a pre-present or a pre-existent cosmic space.

To recap.

Space or better still, cosmic space has to exist first before physical matter can take birth inside it. Cosmic space therefore, is the womb where the physical matter is seeded, weeded, feeded and developed plus matured by the creator, namely god or whatever.

What has been said above can be paraphrased as follows :-

Space or rather cosmic space is the creator or the progenitor or the begetter or the source or the author or the maker of the physical matter of the cosmos. In fact, physical matter is a condensed form of a segment of the ubiquitous and the infinite field of god's awareness aka cosmic space.

SPACE AND PHYSICAL UNIVERSE, FORMED INVOLUNTARILY INSIDE HUMAN AWARENESS DURING THE LATTER'S DREAM SLEEP STATE

Most human awareness', when they go to sleep at night or whenever, they helplessly or powerlessly or if one prefers, involuntarily experience that their **sleep state** has become divided into two parts namely **"dream sleep state"** and "**deep sleep state**".

DREAM SLEEP STATE OF HUMAN AWARENESS

During **"dream sleep state"**, human awareness helplessly or powerlessly or involuntarily perceives and/or experiences or rather, is made to perceive or forced to perceive and/or experience three **awarenessbal phenomena or outward shows or pretences or illusions or phantasms** which are as described below one by one :-

First of all, human awareness in question, becomes unchained or unshackled or unyoked or if it is preferred, becomes free or liberated of the cognizance or the knowledge or the perception or the appreciation of its physical body of which it was and is helplessly or powerlessly or perforce aware or cognizant during each and every spell or bout of its **wakeful state**. Thus, in this state i.e. **dream sleep state** human awareness is forced or compelled to experience or taste or savour **bodiless-ness or disembodied-ness or disbodied-ness. Bodiless-ness or disembodied-ness or disbodied-ness** is the usual state or better still, is the innate or the inherent state of the awareness of the source or the author or the creator of the **physical universe;** the **physical universe** which is helplessly or powerlessly or perforce experienced or tasted or savoured by the human awareness during each and every spell or bout of its **wakeful state** when it helplessly or powerlessly or perforce is made aware or cognizant of its **physical body** as well.

Bodiless or disembodied or disbodied human awareness - which is extant or present during each and every spell or bout of **dream sleep state** - helplessly or powerlessly or perforce perceives the presence or the existence of a **space** inside its own **bodiless or disembodied or disbodied** awareness of the **dream sleep state**. That is to say, it helplessly or powerlessly or perforce perceives the presence or the existence of an **"awarenessbal space"** inside its own **bodiless or disembodied or disbodied** awareness of the **dream sleep state**. This **space** or better still, this **"awarenessbal space"** - which is extant or present inside **the bodiless or disembodied or disbodied** human awareness , during latter's each and every bout or spell of **dream sleep state** - looks or seems or appears to this **bodiless or disembodied or disbodied** human awareness of the **dream sleep state,** one hundred percent similar or one hundred percent identical in every detail to the **space** of the **wakeful state** or better still, to the **"cosmic space"** of the **wakeful state** or more to the point, to the **"awarenessbal space"** of the source or the author or the creator of the **physical universe** which the human awareness in question perceives and experiences perforce or powerlessly or helplessly during each and every bout or spell of its **wakeful state** when it, perforce or powerlessly or helplessly, is also made to become **embodied** once again by its source or author or creator, namely god or whatever.

Finally, inside the above mentioned space or better

still, inside the above mentioned "awarenessbal space" or if one prefers, inside the above mentioned "cosmic space", - formed or developed involuntarily or perforce inside the bodiless or disembodied human awareness of each and every bout or spell of dream sleep state - , the former, namely the bodiless or disembodied human awareness of the dream sleep state, perforce or powerlessly also perceives and experiences a physical universe which is one hundred percent identical in every detail to the physical universe which it perforce or powerlessly perceives and experience, during each and every bout or spell of its wakeful state.

FINAL CONCLUSION REGARDING THE DREAM SLEEP STATE AND ITS TRUE SIGNIFICANCE VIS-A-VIS THE HUMAN AWARENESS

The involuntary experience of the d**ream sleep state** is afforded to human awareness by its **author or** c**reator** in order to remind human awareness of the fact that the perception and the experience of the **wakeful state** on its part is as involuntary as the perception and the experience of the dream sleep state.

Furthermore, the innate or the intrinsic nature of the "***physical universe***", which is perceived and experienced by the human awareness during each and every bout of its wakeful state, is as **_phantasmal or phenomenal or dreamal or dreamlike_** as that of the "**_physical universe_**" which it perceives during each and every bout of its **_dream sleep state._** And the source or the author or the creator of both these **_"universes"_** is the one and the same incredible being who is immortal and aware. It can be called by any name one fancies including the name god.

Therefore, every activity which takes place during each and every bout or spell of the wakeful state of human awareness is involuntary or perforce. That is to say, no activity is voluntary or willed on anyone's part in each and every bout or spell of the wakeful state of human awareness. So, the word or the term "voluntary" should be thrown into the trash bin.

SCIENCE'S VIEW OF THE SOURCE OF THE UNIVERSE

Science's view regarding the source of the 4-D or the four- dimensional spacetime & physical universe revolves round their very famous physical theory called the Theory of Big Bang and Singularity.

THEORY OF BING BANG and SINGULARITY

Cosmic micro wave background radiation (CMBR) was discovered by Penzias & Wilson in 1964. They both received Nobel prize for this in 1978. It is this cosmic microwave background radiation which is put forward or presented by scientists as the indirect physical proof of the occurrence of the event called

Big Bang.

Scientist's narrative is that, this event of Big Bang occurred 13.7 billion light years ago inside the "Starting Point "of the 4-D or the four dimensional spacetime and physical universe. This event then slowly, through the process of "inflation" of this "Starting Point" (as a consequence of Big Bang), led to the formation of the present-day 4-D or the four-dimensional spacetime and physical universe, out of the "physical substance" of this incredible "Starting Point" of physical existence, under the careful direction of the physical law of quantum mechanics, physical law of gravity and the biological law of Darwinian evolution. Scientists afford the name of **"Singularity" or "Cosmic Egg"** to this "Starting Point" of "physical existence".

"Singularity" or the "Starting Point" of "physical existence", that is to say, the "Starting Point" of 4-D or four dimensional spacetime and physical universe is hypothesised by the physical scientists as an **"unconscious physical thing"** made or composed of unconscious physical matter. Scientists state that this **"unconscious physical Singularity"**, made or composed of unconscious physical matter, is the mother or the progenitor or the source or the author or the first cause of man's consciousness just as it also is the mother or the progenitor or the source or author or the first cause of man's physical body as well as the rest of physical universe , not to mention the 4-D or the four dimensional spacetime . In other words, scientist's mathematical, theoretical and empirical or experimental deduction and conclusion is that the mother or the progenitor or the source or the author or the first cause of the extant or the existent consciousness in the universe, for example, human consciousness in the universe, is the unconscious physical matter of *"Singularity"*.

By the way, for obvious reasons, physical scientists will never be able to approach directly, the *"original, unconscious or insentient and physical matter composed Singularity"* in order to verify first-hand its actual presence or existence. Therefore, by necessity, they have to resort to indirect means such as their mathematical and theoretical models plus their experimental results and empirical evidences to define the characteristics of this original, physical *"Singularity"*. This effort of theirs has resulted in defining this original, physical *"Singularity"* as an *"infinitely hot, infinitely dense and infinitesimally small"* thing, composed of physical matter.

Above, one has stated that "For obvious reasons, physical scientists will never be able to approach directly the *original, unconscious or insentient and physical matter composed Singularity* in order to verify first-hand, its actual presence or existence". This is so, firstly, because the original, physical *"Singularity"* existed as such 13.7 billion light years ago only, and therefore, it does not exist now in its original form as **"Singularity".** At the present moment though, it does exist but exists only in its metamorphosed or transmuted form or if one prefers, only in its reconstructed or reconfigured form as the present-day 4-D or four- dimensional- spacetime and physical universe.

The second reason, on account of which scientists can never directly or first-hand approach or reach out to **"original Singularity"** is that at the altitude or level of **"original Singularity"**, nothing else will or can exist apart from the **"original Singularity"** itself.

In short, only physical assets scientists possess in order to indirectly calculate all the above mentioned, three infinities containing attributes of the **"original Singularity"** are the post Big Bang cosmic microwave background radiations (CMBR) plus all their mathematically, theoretically, experimentally and empirically gathered, measured and calculated physical laws of science such as the law of quantum mechanics, the law of gravity and the like.

By the way, at the altitude or the level of **"Singularity"** **or "Cosmic Egg,"** all the laws of Mother Nature such as the law of quantum mechanics, the law of gravity and the like, which form the backbone or better still, which form the foundation of modern physics, breakdown completely. That is to say, all these foundational laws of Mother Nature are absolutely absent or non-existent at the altitude or the level of *"Singularity" or "Cosmic Egg".*

CONCLUSION REGARDING THE NATURE OF SINGULARITY OR COSMIC EGG FROM THE PERSPECTIVE OF AN ADWAITIN

Since it will never be possible for the physical scientists to approach directly or first-hand the *"original Singularity" or "Cosmic Egg"* in order to verify its actual presence or existence first-hand , let alone define its three infinities containing attributes , it will not be unfair on an Adwaitin's part to say that the physical scientists can never be one hundred percent certain in their contention or submission or position that this *"original Singularity"* of their's , absolutely or one hundred percent, was an *"unconscious or insentient physical or material thing made of unconscious or insentient physical matter and not a CONSCIOUS BEING OR A CONSCIOUS REALITY OR A CONSCIOUS TRUTH of intelligence, awareness and emotion called god or whatever".*

~*~*~*~*~

PHYSICAL THEORY OF BIG BANG VS AWARENESSBAL THEORY OF ADWAIT-VEDANTA

SUMMARY

Material scientist's 'singularity' of Big Bang theory fame, as a consequence of being theorised by the former as an absolutely physical entity, will of necessity need a pre-existent cosmic space for its own spatial placement and existence. Therefore, it cannot be the creator of the current cosmic space or, for that matter, creator of the current human awareness' and physical universe.

INTRODUCTION

The Universe exists.

Mankind wonders what or who is its true source?

There are two options for mankind to choose from with regards to the answer of the above question. These are the following :-

One hundred percent physical and therefore one hundred percent insentient "singularity" or "cosmic egg" of Big Bang theory fame.

One hundred percent non-physical, timeless, bodiless, dimensionless awareness of infinite intelligence and emotion called immortal creator of the current universe or god or whatever. This immortal creator of the current universe can be called by any name one fancies. It does not matter.

THERE IS NO ABSOLUTE AND DIRECT PROOF ANYWHERE OF EITHER SINGULARITY OR GOD

The absolute and direct proof of existence of either singularity or god which can be universally accessed by man, does not exist anywhere in the current universe.

What does exist in the current universe regarding both is circumstantial or inferential proof only and nothing else. It is therefore, left to an individual to weigh up these circumstantial proofs and decide which one meets his or her approval or favour.

FIRST OPTION (SINGULARITY)

The first option given above, namely the one hundred percent physical and therefore one hundred percent insentient 'singularity' or 'cosmic egg' of Big Bang theory fame is the choice of the present-day material scientists.

The one hundred percent physical and therefore one hundred percent insentient singularity-centric option of the present-day material scientists with regards to the ultimate source of the current universe, inevitably forces or compels many amongst the vast sea of humanity to think, believe, surmise or conclude that the "Homo sapiens" or the "wise man" of today or better still, the awareness, intelligence and emotion-endowed "Homo sapiens" or "wise man" of today has the status of mere "animated-dirt", mere "animated-physical-matter" or mere "animated-mineral" whose all-important three attributes namely, awareness, intelligence and emotion are mere effluents or excretions, or if it one prefers, are mere by-products or waste-products of the absolutely insentient, physical-matter-composed-entity called 'singularity' or 'cosmic egg' of Big Bang theory fame.

What has been said above can be put in another way.

The one hundred percent physical and therefore absolutely insentient singularity-centric scientific option of the present-day material scientists with regards to the ultimate source of the present-day universe, inevitably forces or compels many amongst the vast sea of humanity to think, believe, surmise or conclude that the "Homo sapiens" or the "wise man" of today or better still, the awareness, intelligence and emotion-endowed "Homo sapiens" of today has the status of mere "animated-dirt", "animated-physical-matter" or "animated-mineral"; whose all important, three attributes namely, awareness, intelligence and emotion are mere effluents, excretions or, if one prefers, mere by-products or waste-products of the absolutely insentient, physical-matter-composed-entity called singularity or cosmic egg of Big Bang theory fame; the awareness, intelligence and emotion which make "Homo sapiens" or "wise man" of today what he truly is, namely, "wise and thoughtful", "analytical and discerning", "compassionate and generous" as well as distinguish, differentiate or demarcate him from the one hundred percent insentient or insensate physical matter.

Those human beings who are the protagonists, supporters, backers or standard bearers of the first option i.e. 'singularity', reject completely the second option, namely the one hundred percent non-physical, timeless, bodiless, dimensionless awareness of infinite intelligence and emotion called immortal creator of the current universe or god or whatever.

INDIRECT OR CIRCUMSTANTIAL EVIDENCE OF BIG BANG THEORY

A. INTRO

Big Bang theory (along with its heart namely 'singularity') is theorized by material scientists as being the source of the current universe; the current universe which consists not only of physical matter but also of non-physical cosmic space and non-physical human awarenesses'.

The indirect or circumstantial evidence of Big Bang theory is described below.

B. INFERENTIAL EVIDENCE OF BIG BANG THEORY

First of all, universe definitely had a beginning. It began about 13.7 billion light years ago. It did not have existence before this time i.e. cosmic time. Therefore, universe will also have an end, an end which will be synchronous with the end of current cosmic time. The end of the current cosmic time nobody knows when or how it will be.

Secondly, as per 'Cosmological-Redshift', universe is expanding i.e. galaxies are moving away from each other at speed which is in proportion to their distance from the observer. Therefore, it can be assumed that in the past, universe was "condensed" or "compacted" or "compressed" into a "single point" which is given the name of "singularity" by material scientists. Thus, current universe was nothing like it is today i.e. expanding.

Thirdly, as per Big Bang theory, "singularity" was an infinitely hot entity 13.7 billion light years ago. This prediction of Big Bang theory is confirmed by the physical presence in the observable universe of cosmic microwave background radiation (CMBR) which pervades the entire universe of today. This CMBR is the residual or the remnant or the leftover thermal radiation or heat from the very early stages of birth of the current universe out of the physical substance of infinitely hot "singularity".

Lastly, the presence in great abundance in the observable universe of hydrogen and helium elements which are the lightest amongst all of the 118 elements of the current universe (both natural and synthesized).

C. *THEORETICAL NATURE OF SINGULARITY AS PER BIG BANG THEORY*

According to Big Bang theory, prior to the "beginning of the current universe", there was absolutely nothing.

However, during the "beginning of the current universe" as well as after the "beginning of the current universe" there was something, namely, the current universe.

Thus, as per Big Bang theory, current universe came into being from "absolutely nothing".

Current universe first arose out of "absolutely nothing" as an entity called "singularity" 13.7 billion light years ago.

This "singularity", over the past 13.7 billion light years, then gradually transformed itself into the current universe under the direction of such immutable laws of Mother Nature as gravity, quantum mechanics and the like on one hand and the Darwinian law of evolution on the other.

What was the true nature of "singularity" 13.7 billion light years ago and how and wherefrom it came into being 13.7 billion light years ago is unknown and most probably is unknowable as per Big Bang theory.

However, Big Bang theory does describe "singularity"

as being an "infinitely hot", "infinitely dense" and "infinitesimally small" physical thing.

On the basis of the above described three cardinal physical characteristics of "singularity" namely, its "infinitely high temperature", "infinitely high density" and "infinitesimally small size", it will not be improper on one's part to assume that this "singularity", 13.7 billion light years ago was some kind of "physical thing" and certainly it was not some kind of "aware thing" such as god or whatever.

Big Bang theory does not say, why "singularity" appeared 13.7 billion light years ago. That is to say, what or who prompted "singularity" to appear as "infinitely hot", "infinitely dense" and "infinitesimally small" thing, out of absolutely nothing, 13.7 billion light years ago.

After the sudden appearance of "singularity", 13.7 billion light years ago, the subsequent events which occurred inside the substance of "singularity" (? physical substance of "singularity") is as follows as per Big Bang theory :-

Subsequent to its sudden appearance 13.7 billion light years ago, the "infinitesimally small in size" substance of "singularity" gradually inflated, expanded, distended or dilated in the manner of a balloon and also gradually began to cool down from an infinitely hot state to the size and temperature of the current universe.

Thus, the entire current universe which includes current "cosmic space", current "human awarenesses" and current "physical matter/physical energy duo", have all emerged out of the physical substance of this incredible "Singularity".

To sum up.

As per Big Bang theory, all the contents of the current expanding universe; contents which include not only the physical matter and physical energy of the current universe but also its such non-physical entities as "cosmic space" and "human awarenesses"; are all the progenies or offspring of an amazing "infinitesimally small sized physical singularity" which 13.7 billion light years ago, suddenly appeared out of nothing for some unknown reason.

The above is the summary of Big Bang theory.

D. THEORETICAL NATURE OF THE EVENT CALLED "BIG BANG" AS PER BIG BANG THEORY

Even though there is this phrase "Big Bang" (thanks to the astronomer and cosmologist Fred Hoyle) in the name "Big Bang theory", there is no explosion in Big Bang theory of the type one associates with a exploding bomb or whatever. In fact, in this theory there is no explosion of any kind whatsoever.

However, there was and there still continues to be an expansion, distention, dilation or inflation of the physical substance of "singularity" in the manner of a "expanding balloon" i.e. an infinitesimally small balloon expanding to the size of the current balloon namely the current cosmic space.

As per Big Bang theory, all galaxies of the current universe are sticking to the outer surface of this balloon. Nobody knows what is happening on the inner surface of this balloon or inside the balloon. Furthermore, nobody knows where this balloon is hanging i.e. is it hanging in some kind of space or cyber space or "mind space" or "awarenessbal space" of someone or something.

Prior to the event of Big Bang, the entity which is called "space" today or better still, which is called "cosmic space" today did not exist. As per Big Bang theory, "space" had a finite beginning which began with the advent of "physical matter" in the universe.

As per Big Bang theory, "singularity" did not appear in "space".

Instead, "space" appeared in "singularity".

The source of the extant human awareness in the current universe or for that matter, the source of any kind of extant awareness in the current universe is not explained in Big Bang theory. In fact, the presence of human awareness, let alone its importance, nay, the presence of any kind of awareness in the current universe and its importance is totally overlooked in Big Bang theory. Presumably, this theory implies without saying so that human awareness or for that matter any awareness in the current universe is the effluent, excretion, by-product or waste-product of presumed physical matter of "Singularity".

Prior to "singularity", nothing existed, not space, not time, not physical matter, not physical energy, not awareness - nothing.

So, the million dollar question is :-

"Where and in what did SINGULARITY appear if not in space?

Big Bang theory does not know the answer to the above question.

INDIRECT OR CIRCUMSTANTIAL EVIDENCE OF GOD

The indirect or the circumstantial evidence for the existence of absolutely non-physical, timeless, bodiless, dimensionless awareness of infinite intelligence and emotion called the creator of current universe or god or whatever, comprises or consists of the presence in the current universe of three extremely important attributes namely awareness, intelligence and emotion, for example awareness, intelligence and emotion of human beings.

It is the awareness of human beings which is the fountain-head, font or source of both intelligence and emotion in human beings.

Another vital point with regards to awareness of human beings is that it is the one and only entity in the current universe which innately is one hundred percent, non-physical or non-material in nature. The rest of the items of the current universe are all, without exception, innately and absolutely, physical or material in nature and nothing else.

By virtue of being the one and only entity in the current universe which innately is one hundred percent non-physical or non-material in nature, human awareness is also the one and only entity in the current universe which innately is absolutely non-dimensional or dimensionless in nature.

On account of being the one and only entity in the current universe which, as a genre or class is innately and absolutely non-dimensional or dimensionless in nature, awareness is and will be the one and only entity anytime, anywhere which does not and will not need or require the pre-presence or pre-existence of cosmic space for its spatial placement and existence.

By using the phrase 'anytime, anywhere' above one means "both before and after the beginning of the current cosmos and the current cosmic time."

What has been said above, can be expressed in another way.

Awareness, as a genre or class is the one and only entity, anytime, anywhere which can and which will exist, if need be, without the pre-presence or pre-existence of cosmic space and thus, is the one and only entity anytime, anywhere which is and which will be in a position to create, make or beget cosmic space, if it wants to.

What has been said above can be put in another way.

Awareness is the one and only entity anytime, anywhere which can be, which will be and in fact which is the absolutely genuine creator, begetter or progenitor of cosmic space.

To recap.

Awareness is the one and only entity anywhere, anytime which can, which will and which does exist in the total absence of cosmic space and hence, which can be, which will be and in fact which is the creator, begetter or progenitor of the current cosmic space.

As said previously, this unique ability of awareness, namely the ability to exist, if need be, in the total absence of cosmic space, is due to its unique essence namely that it is innately and absolutely non-dimensional or dimensionless in characteristic.

The innately and absolutely, non-physical, timeless, bodiless and dimensionless awareness of infinite intelligence and emotion called the creator of current universe or god or whatever therefore, by virtue of being uniquely, innately and absolutely an incredible, dimensionless awareness or more to the point, by virtue of being uniquely, innately and absolutely an incredible, timeless, bodiless and dimensionless awareness, can and does exist eternally without the need or requirement of cosmic space for its own spatial placement and existence.

This unique ability, namely the ability to exist eternally without the need or requirement of cosmic space is not within the reach, scope or capacity of any physical and therefore dimensional being, irrespective of who they are, what they are and where they are.

That is to say, this unique ability to exist eternally without the need or requirement of cosmic space is not within the reach, scope or capacity of all those beings which are physically resident in the current universe, because all these physically resident beings in the current universe are either absolutely physical beings, for example, planets, stars, galaxies, electrons, protons, neutrons and the like or, are part physical, part awarenessbal beings, for example human beings and the like.

As a result, they all without exception, are innately and absolutely dimensional beings, irrespective of whether they are such gargantuan, astral beings as planets, stars, galaxies and the like or, such infinitesimal, sub-atomic beings as electrons, protons, neutrons and the like or, are such part physical, part awarenessbal beings as human beings and the like.

The ability to exist without the need or requirement of the pre-present or pre-existent cosmic space is also not within the reach, scope or capacity of such part physical, part awarenessbal beings (synonym:- embodied aware beings) as human beings and the like who, even though are blessed with a dimensionless awareness, the fact remains that their dimensionless awareness is parked, lodged, positioned or stationed inside a dimensional physical body. As a result, they too cannot exist without the pre-presence or pre-existence of cosmic space for their spatial placement and existence.

On the basis of the above described characteristic of human beings i.e. that they are part physical, part awarenessbal, they are called "fusion-beings".

What has been said above with regards to human beings can be expressed in another way.

Human beings are called "fusion-beings" because they are part physical, part awarenessbal or, part physical, part non-physical or, better still, part dimensional, part non-dimensional or, part dimensional, part dimensionless.

It is obvious that the non-physical cum non-dimensional or, non-physical cum dimensionless part of human beings is human awareness whereas physical cum dimensional part of human beings is human body.

In contrast to human beings, the creator of current universe or god or whatever is the only one of its kind, absolutely non-physical cum non-dimensional or, absolutely non-physical cum dimensionless being.

As said before, in total contrast to the creator of current universe called god or whatever, the physical matter of current universe, as its name implies, is an absolutely physical cum dimensional being.

Thus, from Adwait-Vedantic standpoint, there are three kinds of beings:-

Absolutely non-physical cum non-dimensional being i.e. god.

Absolutely physical cum dimensional being i.e. physical matter/physical energy duo.

"Fusion beings" i.e. human beings.

It has been stated above that the creator of current universe called god or whatever is the only one of its kind, absolutely non-physical cum non-dimensional being. This is the case with the creator of the current universe, on account of the fact that it is the only being anywhere which is absolutely bodiless and dimensionless in essence or more to the point, it is the only being which is absolutely bodiless, dimensionless awareness of infinite intelligence and emotion or, the only being which is absolutely disembodied-cum-dimensionless awareness of infinite intelligence and emotion.

There is no other being anywhere which in identical to this timeless, bodiless, dimensionless awareness of infinite intelligence and emotion called creator of the current universe or god or whatever. That is why it is referred to as being the only one of its kind or unique.

Now let us talk about the common-'o'-garden human awareness.

Since one hundred percent non-physical and therefore non-dimensional or dimensionless human awareness is parked, lodged, positioned or stationed inside a physical cum dimensional body, it is designated as 'embodied awareness'.

The above described characteristic of human awareness, namely that it is an embodied awareness, stands in stark contrast to the awareness of the creator of current universe which, even though is non-physical and therefore non-dimensional or dimensionless in essence in the way of the human awareness, but it is uniquely not parked, lodged, positioned or stationed inside a physical cum dimensional body in the manner of human awareness.

Therefore, the awareness of the creator of current universe called god or whatever is designated as the unique or the only one of its kind, 'disembodied awareness', in contradistinction to man's common-'o'-garden 'embodied awareness'.

The synonymous term for the word 'disembodied' is 'bodiless'. Therefore, god's dimensionless awareness is also described as the only one of its kind or unique, 'bodiless awareness' or, more to the point, as the only one of its kind or unique, 'timeless, bodiless and dimensionless awareness'. God's 'timeless, bodiless, dimensionless awareness' is labeled as being 'timeless' because god's 'bodiless, dimensionless awareness' is immortal or eternal.

By the way, the 'embodied, dimensionless

awareness' of man is as timeless or immortal as the 'disembodied, dimensionless' awareness of the creator of current universe is.

Now let's talk about the 'dimensional physical body' of man inside which the 'embodied, dimensionless awareness' of man is parked, lodged, station or positioned.

The 'dimensional physical body' of man is time-bound or mortal in total contrast to 'dimensionless awareness' of man which is timeless or immortal in the manner of 'dimensionless awareness' of the creator of current universe called god or whatever.

The 'dimensional physical body' of man is time-bound or mortal in the way all the remaining 'dimensional physical bodies' of the current universe are.

Man's time-bound or mortal 'dimensional physical body' is the entity inside which man's 'timeless, dimensionless awareness' is imprisoned, incarcerated, interned or embodied or, if it is preferred, inside which man's 'timeless, dimensionless awareness' is parked, lodged, stationed or positioned.

The timelessness or immortality is the innate nature or essence of man's 'embodied, dimensionless awareness' whereas mortality or temporality is the innate nature or essence of man's 'dimensional physical body'.

The above can be put in another way.

The timelessness or immortality is the innate nature or essence of man's 'dimensionless awareness' which is imprisoned, incarcerated, interned or embodied or, if it is preferred, which is parked, lodged, stationed or positioned inside a time-bound or mortal 'dimensional physical body'.

The timelessness or immortality is the innate nature or essence of man's 'embodied, dimensionless awareness' on account of the fact that it is an 'original, authentic, pure or pristine' piece, portion, section or segment of god's 'timeless, bodiless, dimensionless awareness', nothing more nothing less.

In contrast, the mortality or temporality is the innate nature or essence of man's 'dimensional physical body' on account of the fact it is not an 'original, authentic, pure or pristine' piece, portion, section or segment of god's 'timeless, bodiless, dimensionless awareness'. Instead, it is a 'condensed, compacted or compressed' piece, portion, section or segment of god's 'timeless, bodiless, dimensionless awareness', nothing more nothing less.

To recap.

It is man's 'dimensional physical body' only which is time-bound or mortal and not man's 'dimensionless awareness' which is as timeless or immortal as god's 'timeless, bodiless, dimensionless awareness' is.

To reiterate.

Man's time-bound or mortal, 'dimensional physical body' is the entity inside which man's 'timeless, dimensionless awareness' is imprisoned, incarcerated, interned or embodied or, if it is preferred, inside which man's 'timeless, dimensionless awareness' is parked, lodged, stationed or positioned.

Man's 'embodied, dimensionless awareness' is as timeless or immortal as the 'disembodied, dimensionless awareness' of the creator of current universe is. This is the case on account of the fact that man's 'embodied, dimensionless awareness' is an 'original, authentic, pure or pristine' piece, portion, section or segment of god's 'timeless, bodiless, dimensionless awareness', nothing more nothing less.

In contrast, mortality or temporality is the innate nature or essence of man's 'dimensional physical body' on account of the fact it is not an 'original, authentic, pure or pristine' piece, portion, section or segment of god's 'timeless, bodiless, dimensionless awareness'. Instead, it is a 'condensed, compacted or compressed' piece, portion, section or segment of god's 'timeless, bodiless dimensionless awareness', nothing more nothing less.

It will be useful to remind oneself once again that it is

only the 'dimensional physical body' of man which is time-bound or mortal just as are all the other dimensional physical bodies of the current universe.

As said before, man's time-bound or mortal, 'dimensional physical body' is the entity inside which man's 'timeless, dimensionless awareness' is embodied, imprisoned, incarcerated or interned or, if it is preferred, is parked, lodged, stationed or positioned.

It is the dimensional silhouette or configuration of an entity in the current universe which makes that entity 'time-bound and space-bound' plus bestows upon that entity all the characteristics as well as the epithet of being 'physical' in essence.

'Dimensionless-ness' or 'non-dimensional-ness' is the hallmark of all the awareness'. Associated with the 'universal dimensionless-ness' or 'universal non-dimensional-ness' of all the awareness' is the 'universal timeless-ness' and 'universal non-spatial-ness' of all the awareness'.

Thus, 'dimensionless-ness' or 'non-dimensional-ness', 'non-spatial-ness' and 'timeless-ness' are the three cardinal or fundamental characteristics of all the awareness', irrespective of whether the awareness in question is the awareness of the creator of current universe or the awareness of man.

In absolute contrast to the 'universal dimensionless-ness' or 'non-dimensional-ness' of all the awareness', 'universal dimensional-ness' is the hallmark of all the physical objects of the current universe.

Associated with the 'universal dimensional-ness' of all the physical objects of the current universe, is the 'universal spatial-ness' plus the 'universal temporality' of all the physical objects of the current universe.

Thus, 'dimensional-ness', 'spatial-ness' and 'temporality' are the three cardinal or fundamental characteristics of all the physical objects of the current universe, irrespective of whether the physical objects in question are the physical bodies of human beings, or, the physical bodies of such astral objects in the current universe as planets, star, galaxies and the like or, the physical bodies of such subatomic particles in the current universe as electrons, protons, neutrons and the like.

At this juncture, one more vital point regarding 'timeless, dimensionless, embodied human awareness' requires to be discussed.

On account of its 'embodied-ness' or 'embodied-persona' or, on account of its imprisonment, incarceration or internment inside a 'time-bound and dimensional' physical body, 'timeless, dimensionless' human awareness cannot exist in the absence of a pre-present or pre-existent cosmic space.

'Timeless, dimensionless, embodied human awareness', perforce or of necessity, needs or requires a pre-present or pre-existent cosmic space for the spatial placement & existence of its 'time-bound and dimensional physical body'. The 'time-bound and dimensional physical body' is the entity inside which 'timeless, dimensionless human awareness' is embodied, imprisoned, incarcerated or interned or, is parked, lodged, stationed or positioned.

In other words, the unavoidable need or requirement on the part of 'dimensionless, timeless human awareness' for a pre-present or pre-existent cosmic space is on account of the fact that its 'prison' or its 'dungeon' namely its 'dimensional and ephemeral' or 'dimensional and time-bound' physical body needs or requires a pre-present or pre-existent cosmic space for its spatial placement and existence.

'Dimensionless, timeless human awareness' itself does not need or require a pre-present or pre-existent cosmic space for its spatial placement and existence because it is 'dimensionless' in characteristic in the way god's awareness is.

What has been said above can be put in another way.

In the current universe, the existence of

'dimensionless, timeless human awareness' is not possible and therefore cannot take place without it being first imprisoned, incarcerated, interned or embodied or, if one prefers, without it being first parked, lodged, stationed or positioned inside a 'dimensional and ephemeral' or a 'dimensional and time-bound' physical body. This 'dimensional and ephemeral' or 'dimensional and time-bound' physical body is a sine qua non or a necessary pre-condition for the existence of 'timeless, dimensionless human awareness' in the current 'dimensional universe' or better still, in the current

'3-D or three-dimensional universe.

To recap.

Unlike god's 'timeless, dimensionless awareness' which is uniquely 'disembodied or bodiless' and therefore can and does exist without the need or requirement of a pre-present or pre-existent cosmic space, man's 'timeless, dimensionless awareness', on the other hand, is embodied, imprisoned, incarcerated or interned inside a 'dimensional and time-bound physical body' and therefore, it cannot exist in the absence of a pre-present or pre-existent cosmic space. This is the case with regards to human awareness despite the fact that it itself is innately and absolutely 'non-physical & non-dimensional' or ' non-physical and dimensionless' in nature in the manner of the 'non-physical and non-dimensional' or 'non-physical and dimensionless awareness' of the creator of current universe. The reason for this is that the physical body of man inside which human awareness is parked, lodged, stationed or positioned or, embodied, imprisoned, incarcerated or interned, is a dimensional entity and therefore for its spatial placement and existence, a pre-present or pre-present dimensional cosmic space is a must or a sine qua non or an indispensable pre-condition.

As said before, in total contrast to common-'o'-garden, 'non-physical and non-dimensional' or 'non-physical and dimensionless', embodied awareness of human beings, god's 'non-physical and non-dimensional' or 'non-physical and dimensionless' awareness, is the only one of its kind or unique, 'disembodied or bodiless' awareness.

AWARENESS

It will be worthwhile to re-state here the following facts about all the awareness', irrespective of whether it is the only one of its kind or unique disembodied or bodiless awareness of the creator of current universe called god or whatever or they are the common-'o'-garden embodied awareness' of human beings.

All awareness', without exception, are innately and absolutely non-physical, timeless, dimensionless beings which are endowed not only with sentience and intelligence but emotions too. Sentience is the ability to feel, perceive, experience, think, reason and be aware of oneself as an aware or conscious being subjectively.

There is another universal characteristic of all awareness'. Since it is a universal characteristic of all awareness', it is possessed by all awareness', irrespective of whether it is the only one of its kind or unique disembodied or bodiless awareness of the creator of current universe called god or whatever or, it is the common-'o'-garden embodied awareness of human beings.

This universal characteristic of all awareness' is described below.

Irrespective of whether an awareness in question is the only one of its kind or unique disembodied or bodiless awareness of the creator of current universe called god or whatever or, it is the common-'o'-garden embodied awareness of a human being, it is the only entity anytime, anywhere which is innately empowered to give rise, genesis, birth or nascence to a space or, more to the point, to an awarenessbal space inside itself.

A space or an awarenessbal space, which an awareness gives rise, genesis, birth or nascence, exists inside the awareness in itself and nowhere else.

It will be worthwhile to repeat here once again that a space or, more to the point, an awarenessbal space which an awareness gives rise, genesis, birth or nascence, always exists inside the awareness and nowhere else.

A space or an awarenessbal space which an awareness gives rise, genesis, birth or nascence, inside itself can be called by any name one fancies including the name cosmic space.

Earlier it has been stated that all awareness' are 'dimensionless' beings.

The meaning of being 'dimensionless' in characteristic is that the 'beings' in question namely all the awareness' themselves, unlike the physical matter of the current universe, do not need or require any space or more precisely, the cosmic space for their spatial placement and existence. They therefore, can and do exist both in the absence and the presence of cosmic space.

SPACE

Any awareness can give birth or nascence to space or, more accurately, to awarenessbal-space inside itself anytime it wants.

A space or an awarenessbal-space can be given birth or nascence by any awareness inside itself through the process of expansion, distention, dilation or inflation of itself and nothing else.

Both the source and place of birth and existence of a space or an awarenessbal-space will always be an awareness and awareness only and nothing else, irrespective of whether the space or the awarenessbal-space in question is the current cosmic-space or, a space or an awarenessbal-space which any human awareness can and does give birth or nascence to inside itself during its common-'o'garden, awarenessbal activity, pass time or game called daydreaming or reverieing.

It will be worthwhile to repeat here once again all that has been said above with regards to space or, more to the point, awarenessbal-space because it is extremely important with regards to understanding the true nature of the current cosmic-space and the true nature of the current physical universe.

Both the source and place of birth and existence of space or, more to the point, awarenessbal-space can and will always be an awareness and awareness only and nothing else, irrespective of whether the awareness in question is the awareness of the creator of current universe or the awareness of a human being.

In fact, all awareness', irrespective of whether the awareness in question is the only one of its kind or unique, timeless, dimensionless and disembodied awareness of the creator of current universe called god or whatever or, a common-'o'-garden, timeless, dimensionless and embodied awareness of a human being, can and do give birth or nascence to a space or, more to the point, to an awarenessbal-space inside themselves, (whenever and wherever their mood or fancy dictates them to do so), by the process of volitional or voluntary expansion, distention, dilation or inflation of themselves during their common-'o'-garden or mundane, awarenessbal activity, pass time or game called daydreaming or reverieing. A space or an awarenessbal-space, thus given birth, genesis or nascence by an awareness inside itself through the process of volitional or voluntary expansion, distention, dilation or inflation of itself can be called by any name one fancies including the name cosmic-space.

What name a space or an awarenessbal-space is granted or awarded, does not matter.

However, what does matter though, is that one must always bear in mind the vital fact that any space they see or encounter anytime, anywhere in the current universe, is an awarenessbal-space and nothing but an awarenessbal-space. And it has been given birth or nascence by an awareness and awareness only and by nothing else. Furthermore, it exists inside that awareness and awareness only and nowhere else. Additionally, one must bear in mind that this space or awarenessbal-space has been given birth or nascence by the awareness in question inside itself through the process of volitional or voluntary expansion, distention, dilation or inflation of itself and nothing else during the common-'o'-garden or mundane awarenessbal activity, pass time or game on its part called daydreaming or reverieing.

The above statement vis-a-vis a space or, more to the point, via-a-vis an awarenessbal-space is true, irrespective of whether it is an awarenessbal-space extant inside the only one of its kind or unique timeless, dimensionless and disembodied awareness of the creator of current universe or, it is an awarenessbal-space extant inside the timeless, dimensionless and embodied awareness of a human being.

The implication of what has been said above with regards to a space or, more precisely with regards to an awarenessbal-space is the following.

What is labeled as cosmic-space by human awareness' in their wakeful state and inside which the current physical universe of man's wakeful state is floating, wafting or levitating plus whirling, twirling or spiralling non-stop is nothing but an awarenessbal-space which exists inside the timeless, dimensionless and disembodied awareness of the creator of current universe and nowhere else. This timeless, dimensionless and disembodied awareness of the creator of current universe has given birth or nascence to this awarenessbal-space aka cosmic-space inside itself, through the process of expansion, distention, dilation or inflation of itself and nothing else during its volitional or voluntary, common-'o'-garden or mundane, awarenessbal activity, pass time or game called daydreaming or reverieing.

To reiterate.

The entity labeled as cosmic-space by human awareness' in their wakeful state and inside which the current physical universe of man's wakeful state is floating, wafting or levitating plus whirling, twirling or spiralling non-stop, has been given birth or nascence by the timeless, dimensionless and disembodied awareness of the creator of current universe inside itself through the process of expansion, distention, dilation or inflation of itself and nothing else during its volitional or voluntary awarenessbal pass time, game or sport called daydreaming or reverieing.

Thus, the current cosmic-space is parked, lodged, stationed or positioned inside the timeless, dimensionless and disembodied awareness of the creator of current universe. Therefore, the more appropriate epithet or title for the current cosmic-space will be the awarenessbal-space of the creator of current universe or, even the mind-space of the creator of current universe because the terms 'awareness' and 'mind' are accepted to be identical in the Adwait-Vedantic realm.

To sum up.

The innate nature of any space, always is and always will be that of an awarenessbal-space or mind-space and nothing else, irrespective of whether the space in question is an awarenessbal-space or mind space extant inside the timeless, dimensionless and disembodied awareness or mind of the creator of current universe or an awarenessbal-space or mind-space extant inside the timeless, dimensionless and embodied awareness or mind of a human being.

The above-mentioned truth about the innate nature of any space anywhere, irrespective of its size and situation or expanse and location, can be elaborated in another manner which is as follows.

Any space anywhere encountered or seen by human awareness or mind; irrespective of its size and situation or expanse and location; in terms of absolute truth, always is and always will be an awarenessbal-space or mind-space and nothing else.

Or better still, any space anywhere, anytime encountered or seen by human awareness or mind; irrespective of its size and situation or expanse location; always is and always will be a space which is extant or present inside an awareness and awareness only and nowhere else or inside a mind and mind only and nowhere else.

Thus, it can be stated that all awareness' or minds, without exception are space-creating truths or beings.

The space-creating awareness or mind creates or forms a space or, more to the point, an awarenessbal-space or mind-space inside itself through the process of expansion, distention, dilation or inflation of itself and nothing else during its volitional or voluntary awarenessbal or mental activity, pass time or game called daydreaming or reverieing. And this awareness or mind in question can be the only one of its kind or unique timeless, dimensionless and disembodied awareness or mind of the creator of current universe or the common-'o'-garden timeless, dimensionless and embodied awareness or mind of a human being. The point to emphasise here is that the nature of space in both these instances is the same. That is to say, the nature of space in both these cases is that of an awarenessbal-space or a mind-space, nothing more nothing less.

To repeat.

The method by which either the only one of its kind or unique timeless, dimensionless and disembodied awareness or mind of the creator of current universe or the timeless, dimensionless and embodied awareness or mind of a human being creates a space or, more to the point, an awarenessbal-space or mind-space inside its individual self is exactly the same. This method consists of volitional or voluntary expansion, distention, dilation or inflation, on each of their part, of their individual self, i.e. of their individual awareness or mind during their awarenessbal or mental activity, pass time or game called daydreaming or reveriering, nothing more nothing less.

The motive, the incentive or the lure on the part of awareness or mind of the creator of current universe to create or, give birth or nascence to a space or, more to the point, to an awarenessbal-space or mind-space inside its timeless, dimensionless and disembodied awareness or mind or, more to the point, inside its timeless, dimensionless and disembodied self, by the process of expansion, distention, dilation or inflation of its awareness, mind or the self was, is and always will be to create or give birth or nascence to a day-dream-stuff composed or a reverie-stuff composed universe inside its expanded, distended, dilated or inflated awareness, mind or the self through the instrumentality of daydreaming or reverieing on its part in order to amuse, entertain, regale or titillate itself, nothing more nothing less.

Regretfully, this day-dream-stuff or reverie-stuff composed universe, thus created or given nascence or birth by the creator of current universe inside its expanded, distended, dilated or inflated awareness, mind or the self or, if it is preferred, inside its awarenessbal-space, mind-space or the self, is labeled as physical universe by mankind and creator's awarenessbal-space, mind-space or the self, i.e. creator's expanded, distended, dilated or inflated awareness, mind or the self is labeled as cosmic-space by mankind.

Additionally, the day-dream-stuff composed or reverie-stuff composed universe, created or given nascence or birth by the creator of current universe inside its expanded, distended, dilated or inflated awareness, mind or the self, i.e. inside its awarenessbal-space, mind-space or the self, is wrongly labeled by mankind as physical universe and is accepted by mankind as being absolutely or one hundred percent real. All this misapprehension, error, confusion, mix-up or delusion on mankind's part is on account of the lack of proper knowledge on its part of the true nature of current universe, the true nature of current cosmic-space, the true nature of itself and the true nature of the creator of current universe.

This lack of proper knowledge on mankind's part with regards to the true nature of current universe, the true nature of current cosmic space, the true nature of itself and the true nature of the creator of current universe is quite understandable.

This lack of proper knowledge on mankind's part vis-a-vis the true nature of the current universe, the true nature of the current cosmic space, the true nature of itself and the true nature of the creator of current universe is quite understandable because mankind is not, in the slightest bit, apart from the current universe. On the contrary, it is every bit, an integral part of this day-dream-stuff composed or reverie-stuff composed current universe. In fact, whether mankind has realised or not, it merely is one amongst the countless other items of this incredible day-dream-stuff composed or reverie-stuff composed current universe. Therefore, for all intents and purposes it is impossible for mankind to grasp the truth about itself, about the current cosmic space and about the current universe, let alone the truth about the creator of current universe, the creator of current cosmic-space and the creator of current physical-matter-physical-energy duo.

Now, let's talk about once again about the timeless, dimensionless awareness of human beings or, more precisely, about the timeless, dimensionless true-self of human beings. Here timeless, dimensionless awareness of human beings has been equated or identified with the true-self of human beings because the timeless, dimensionless true-self of human beings is not their time-bound and dimensional physical body - as majority of humanity thinks it is - but their timeless, dimensionless awareness which is absolutely non-physical in nature, in stark contrast to their physical body which is, as its name implies, physical in nature.

Here, one will talk about the timeless, dimensionless true-self of human beings namely, the timeless, dimensionless awareness of human beings, because this talk about the timeless, dimensionless awareness of human beings or, this talk about the timeless, dimensionless true-self of human beings is very apt or fitting in the context of all that has already been said above, with regards to the timeless, dimensionless awareness of the creator of current universe or, the timeless, dimensionless true-self of the creator of current universe. This is the case because at the supreme level, both these beings are two sides of the same coin or, both these beings are very closely related although they seem different at the worldly level.

The timeless, dimensionless awareness of human beings or, the timeless, dimensionless true-self of human beings also creates or gives birth or nascence to space or, more accurately, to awarenessbal-space or mind-space and its associated day-dream-stuff composed or reverie-stuff composed universe inside its timeless, dimensionless awareness or, more precisely, inside its timeless, dimensionless true-self, every time its timeless, dimensionless true-self or every time its timeless, dimensionless awareness volitionally or voluntarily or on its own accord engages in its common-'o'-garden, awarenessbal activity, pastime, game or sport called daydreaming or reverieing in order to amuse, entertain, regale or titillate itself.

Amazingly, this space or better still, this awarenessbal-space or mind-space is also created or, given nascence or birth inside the timeless, dimensionless awareness of human being or, more to the point, inside the timeless, dimensionless true-self of human being, every time this true-self of human beings or, every time this awareness of human beings comes into contact with or runs into its dream-sleep-experience during its each and every sleep-state.

However, one will like to point out that the creation or, giving birth or nascence to space or, more precisely, to awarenessbal-space or mind-space which takes place or comes about inside the timeless, dimensionless awareness of human beings or inside the timeless, dimensionless true-self of human beings during its rendezvous or meeting with dream-sleep-experience, takes place or comes about, involuntarily or compulsorily under the impact or the influence of mighty or omnipotent will of the timeless, dimensionless awareness or the timeless, dimensionless true-self of the creator of current universe and not by the choice of the individual concerned. By the way, may one mention here that this creator of the current universe is also evidently the creator of the timeless, dimensionless awareness or the timeless, dimensionless true-self of human beings.

The reason why the timeless, dimensionless awareness or the timeless, dimensionless true-self of the creator of current universe has bestowed, conferred or vouchsafed upon the timeless, dimensionless awareness or the timeless, dimensionless true-self of human beings, the ability to volitionally daydream on one hand and compulsorily to have rendezvous with the dream-sleep experience (during its each and every sleep-caper or sleep-state) on the other, is to educate the latter about the true or fundamental nature of the current universe which regretfully is believed and accepted by the latter to be absolutely real in comparison to its daydreams on one hand and to its dream-sleep experiences on the other, both of which are regarded by it to be absolutely unreal.

The timeless, dimensionless awareness or the timeless, dimensionless true-self of human beings regretfully regards the current universe to be absolutely real in comparison to its daydreams on one hand and to its dream-sleep experiences on the other (both of which it regards to be absolutely unreal), on account of huge misapprehension, error, confusion, mix-up or delusion on its part with regards to the true nature of the current universe.

To reiterate.

On account of huge misapprehension, error, confusion, mix-up or delusion on its part with regards to the true or fundamental nature of the current universe, the timeless, dimensionless awareness or the timeless, dimensionless true-self of human beings regretfully regards the current universe to be absolutely real in comparison to its daydreams on one hand and its dream-sleep experiences on the other, both of which it regards to be absolutely unreal.

However, the reality is very different. This reality is that the current universe is as much a 'phenomenon' as all the daydreams of human awareness' are, as well as all the dream-sleep experiences of human awareness' are. In other words, all the three truths, presently under discussion, namely the current universe, the daydreams of human awareness', and the dream-sleep experiences of human awareness' are all 'phenomenal realities' only and nothing but 'phenomenal realities' only, nothing more nothing less. There is no fundamental difference between these three truths (i.e. the current universe, the daydreams of human awareness', and the dream-sleep experiences of human beings) vis-a-vis their respective intrinsic nature or essence, despite the mistaken, delusional, confused or wide of the mark belief on the part of the timeless, dimensionless awareness of human beings or the timeless, dimensionless true-self of human beings that the current universe is absolutely real whereas their daydreams as well as their dream-sleep experiences are both absolutely unreal.

The word 'phenomenon' used above has been employed in the following sense :-

"A truth labeled as 'phenomenon' is merely an occurrence, circumstance or fact that is perceptible by the senses or, merely an object of perception and experience by the senses or, merely a thing as it appears and is interpreted in perception and reflection by the senses, as distinguished from its real nature or intrinsic nature as a thing-in-itself or, as a thing-as-it is-in-itself ".

The current universe is quintessentially an 'awarenessbal truth' and nothing but an 'awarenessbal truth'. What is labeled as 'physical truth' in the current universe i.e. matter and energy, is quintessentially 'awarenessbal' in make-up because what is called 'physical' in the current universe is a mere condensed, compacted or compressed form of the timeless, dimensionless awareness of the creator of current universe, nothing more nothing less. A segment of the timeless, dimensionless awareness of the creator of current universe has become condensed, compacted or compressed under the impact or the influence of the mighty or the omnipotent will of the creator of current universe to give rise to physical matter of the current universe on one hand and the physical energy of the current universe on the other.

At the moment of denouement, ending, finale or termination of the current universe; which surely will take place one day sometime in future unknown; both energy and matter will reassume, repossess, recover or regain their awarenessbal form and will once again become absolutely one with the timeless, dimensionless awareness of their creator from whom they temporarily became separated as per latter's mighty or the omnipotent will, in order to become physical energy on one hand and physical matter on the other so as to give rise to variety, diversity or multiplicity which is perceived and experienced today in the otherwise ubiquitous and the infinite field of timeless, dimensionless awareness of the creator of current universe called cosmic space.

Just as steam, water and ice are fundamentally three forms of one and the same thing, each having different sets of characteristics, similarly, the timeless, dimensionless awareness of the creator of current universe on one hand and physical energy and physical matter of the current universe on the other are quintessentially one and the same thing at the fundamental level, despite each having totally different sets of characteristics.

ADWAIT-VEDANTIC REALM

In Adwait-Vedantic realm, the two terms, 'awareness' and 'consciousness' are used interchangeably.

A PITHY DESCRIPTION OF THE CURRENT COSMOS AND ITS   CREATOR IS AS FOLLOWS in Adwait-Vedantic parlance or dialect:-

THE CURRENT COSMOS IS A CONSCIOUSNESSBAL DANCE   OF COSMIC PROPORTION, CREATED CONSCIOUSNESSBALLY BY TIMELESS, DIMENSIONLESS CONSCIOUSNESS OF INFINITE   INTELLIGENCE   AND EMOTION, INSIDE ITS   OWN   CONSCIOUSNESSBAL - SPACE   CALLED, COSMIC SPACE.

THE ABOVE DESCRIPTION OF CURRENT COSMOS AND ITS CREATOR CAN BE EXPRESSED IN ANOTHER FORM :-

A (Creator's Awareness) is equivalent to

 E (Energy) which is equivalent to

M (Matter) and vice versa.

OR

A (Creator's Awareness) $\equiv$ **E** (Energy) $\equiv$ **M**(Matter)

Existence (Sat), under the aegis of its bedrock called Awareness (Chid) is endless. The twin called Existence and Awareness (Sat & Chid) constitute an eternal continuum.

The eternal bedrock of eternal Existence is eternal Awareness. And this eternal bedrock of eternal Existence i.e. eternal Awareness is eternally endowed with infinite intelligence and emotions. It is also eternally endowed with the capacity to transmute a segment of itself, if it so desires, into physical matter on one hand and physical energy on the other by the might of its omnipotent will during its awarenessbal pass time, game or sport called daydreaming.

There never will be the state of non-existence, cipher, zero or shoonya.

DREAM-SLEEP STATE OF HUMAN AWARENESS

Human awareness is unique with regards to one of its attributes.

This attribute consists of the experience on the part of dream-sleep state during each and every sleep-activity, all through its lives.

This special attribute of human awareness i.e. of experiencing dream-sleep state or running into dream-sleep state experience, during each and every sleep-activity, is a special gift of the creator of current universe to human awareness which has been bestowed upon the latter for a very special reason.

The gift of dream-sleep state which all human awareness' are blessed with, is not possessed by **the only one of its kind** or unique awareness of the creator of current universe because the creator of current universe has no need of this special attribute.

The only one of its kind or unique awareness of the creator of current universe has no need for this special attribute namely the attribute of experiencing dream-sleep state or running into the dream-sleep state experience owing to the fact that this creator is what it is. This creator knows what it is, namely, that it is the source, the author, the creator or the progenitor of every being, thing, event and phenomenon of the current universe including all human bodies, all human awareness' and latter's all thoughts, dreams, desires and activities plus propensities or proclivities.

But human awareness' do not know what they truly are and therefore their creator namely the incredible god or whatever, out of its immense compassion towards them, has accorded, awarded, granted, gifted or vouchsafed to them this unique attribute i.e. the attribute of experiencing dream-sleep state or running into the dream-sleep state experience during their each and every sleep-activity so that they i.e. human awareness' one day become cognisant, conscious, aware, apprised or abreast of their true nature, nay, become cognisant, conscious, aware, apprised or abreast of the true nature of the current universe which consists of myriad beings and things of mind-boggling diversity.

As said before, this special attribute called the experience of dream-sleep state which every human awareness rendezvouses or runs into during its each and every sleep-activity, has been vouchsafed to it by its incredible creator for a very special reason. This reason is described below.

Every human awareness makes or create a space or more to the point, an awarenessbal space inside itself whenever it finds itself into its dream-sleep state which materialises mostly during its sleep-activity at night but it can materialise during its sleep-activity at any time.

However here, the most important point all human awareness' must take notice of, is that during their dream-sleep-state, they make or create a space or, more to the point, an awarenessbal space inside their awareness along with its **'spectacular-dream-**

sleep-state-panorama' called **'dreams of dream-sleep state',** not of their own accord or of their own volition or of their own free will, as is the case during their wakeful state's awarenessbal activity, game, sport or pastime called daydreaming but perforce under the purview or under the sphere of influence of the mighty or the omnipotent will of their creator called god or whatever.

Now let's talk about the true nature of the current cosmic space.

God's incredible, timeless, bodiless and dimensionless awareness of infinite intelligence and emotion has created or made inside itself a **space** or, more to the point, an **awarenessbal space** namely the current **cosmic space** along with its **'spectacular daydream-stuff composed panorama'** or **universe** namely the current **physical universe** through the process of expansion, distention, dilation or inflation of its awareness plus its associated, connected or related awarenessbal activity, game, sport or pastime called daydreaming or reverieing.

The creator of current universe has made or created space or more to the point, awarenessbal space namely the current **cosmic space** inside its timeless, bodiless and dimensionless awareness of infinite intelligence and emotion in order to spatially, geographically, geometrically or territorially accommodate its current daydream-stuff composed, spectacular **awarenessbal panorama** called physical universe.

God's current daydream-stuff or reverie-stuff composed, spectacular **awarenessbal panorama** called physical universe, has been created or generated awarenessbally only by god and therefore it exists inside god's awareness only and nowhere else. It has been made or created by god simply as a game, sport or pastime through the wellknown and common-'o'-garden process of daydreaming or reverieing on its part and nothing else, in order to amuse, entertain, regale or titillate itself and nothing else.

Thus, inside god's awarenessbal space namely, the current cosmic space, the present-day spectacular **awarenessbal panorama** called physical universe is floating, wafting or levitating plus whirling, twirling or spiralling non-stop as god's daydream, reverie, imagery, dreamry or phantasy only and nothing else.

God's daydream-stuff or reverie-stuff composed current spectacular **awarenessbal panorama** called physical universe has been floating, wafting or levitating plus whirling, twirling or spiralling non- stop inside god's awarenessbal space namely the current cosmic space since the moment of its inception inside god's awareness 13.7 billion light years ago.

To reiterate.

The current cosmic space is nothing but god's awarenessbal space which is parked, lodged, stationed or positioned inside god's awareness only and nowhere else. In other words, the current cosmic space is an incredible space which is extant inside god's awareness only and nowhere else. And god itself is an awe-inspiring and **the only one of its kind,** timeless, bodiless and dimensionless awareness of infinite intelligence and emotion.

Whenever human beings see a space anywhere they must always remember whatever has been said above with regards to the true nature of the current cosmic space

The above can be put in another way.

Human beings must not lose sight of the fact that the nature of the current, ubiquitous and the infinite field of god's awareness called cosmic space - inside which, the present-day physical universe has been floating, wafting or levitating plus whirling, twirling or spiralling non-stop from the moment of its genesis by god 13.7 billion light years ago - is that of an awarenessbal space or, more precisely, is that of an awarenessbal space of god and nothing else. It is therefore extant inside the awareness of god only and nowhere else.

Furthermore, human awareness must also always remember that god, in its original state i.e. prior to giving birth or nascence to its current ubiquitous and the infinite field of awareness called cosmic space, was an incredible, timeless, bodiless, dimensionless awareness of infinite intelligence and emotion. However, at the present moment though it exists as the ubiquitous and the infinite field of awareness called cosmic space. Each and every human being must be mindful of this extraordinary truth about cosmic space because by knowing this truth about cosmic space he or she will be able to understand the true nature of the current universe of both physical kind as well as of awarenessbal kind.

Thus, human beings now, do not have to imagine that the creator of current universe called god or whatever lives in some faraway, obscure, or very distant in space & time place called heaven, paradise or whatever. This creator or god is everywhere in the form of the ubiquitous and the infinite field of divine awareness called cosmic space. This is an extremely heart-warming, uplifting and soul-stirring truth for mankind to apprehend, imbibe, assimilate and internalise.

To sum up.

The current cosmic space is the awarenessbal space which is extant inside the awareness of the creator of current universe.

In other words, current cosmic space is the mind of creator or the mind of god or, if it is preferred, is the awareness of creator or the awareness of god.

The space or, more to the point, the awarenessbal space can be created by all awareness' including the awareness of god as well as the awareness of all human beings inside their own respective self by the volitional or voluntary process of expansion, distention, dilation or inflation of their own respective self.

The awarenessbal space created by god's awareness inside itself by the volitional or the voluntary process of expansion, distention, dilation or inflation of itself is given the appellation of cosmic space by human awareness'.

The current cosmic space is the entity inside which the current physical universe is floating, wafting or levitating plus whirling, twirling or spiralling non-stop as a mere daydream, reverie, dreamry or imagery of god's awareness, nothing more nothing less.

All that has been said above has been catalogued or set out below in a numbered fashion.

The purpose of all that has been said up till now is to draw the attention of mankind to the supreme truth vis-a-vis the real nature of the current cosmic space and the real nature of the current physical universe.

This supreme truth is unknown to most of mankind.

The enigmatic entity designated as cosmic space by human awareness can be seen by the human eye and yet can't be touched, tasted, smelt or heard plus whose essence or quintessence is non-physical.

Inside this current cosmic space, the current physical universe is floating, wafting or levitating plus whirling, twirling or spiraling non- stop from the moment of its genesis by god, 13.7 billion light years ago.

Such an enigmatic cosmic space is in fact the awarenessbal space or mind space of the timeless, bodiless and dimensionless awareness of infinite intelligence and emotion called the creator of current universe or god or whatever.

This creator, god or whatever of the current universe is the source or author of the present-day physical universe.

This immortal awareness or this immortal source of the present-day physical universe can be called god or anything else one fancies. It does not matter.

All awareness' are non-physical and non-dimensional or dimensionless in their inherent or essential make-up regardless of whether it is the awareness of the creator of current universe or the awareness of human beings.

In stark contrast to awareness, the inherent or essential make-up of physical matter is dimensional, spatial, territorial, geographic, geometric or is related to space and size.

Hence, physical matter, regardless of where it exists - perforce or of necessity, will always need or require a pre-present or pre-existent space, geography, geometry or territory for its own spatial placement and existence, whatever be its size, dimension, measurement or proportion.

That is to say, physical matter - whatever be its size, dimension, measurement or proportion i.e. even if it is of an infinitesimally small size, dimension, measurement or proportion, as the material scientists of the present-day have theorised that their physical singularity of Big Bang theory fame is - will need or require a pre-present or pre-existent space, geography, geometry or territory for its own spatial, territorial, geographic or geometric placement and existent.

Therefore, such a physical singularity of Big Bang theory fame cannot exist in the absence of a pre-present or pre-existent space, geography, geometry or territory despite its infinitesimally small size, dimension, measurement or proportion.

In other words, it is impossible for such a physical thing

as singularity of Big Bang theory fame to be the source, creator, maker or progenitor of the current space or, more to the point, of the current cosmic space despite the theoretical claim of the present-day material scientists.

Space or more to the point, awarenessbal space is a sine qua non or an indispensable pre-condition for the existence of any kind of physical matter including the physical matter of infinitesimally small sized singularity of Big Bang theory fame.

In other words, it is of no consequence whether the physical matter in question is of infinitesimally small size, dimension, measurement or proportion- as the present-day material scientists theorise that their physical singularity of Big Bang theory fame is - or, of colossal size i.e. planets, stars, galaxies and the like, physical matter, perforce or of necessity will always need or require a pre-present or pre-existent space or, more to the point, a pre-present or pre-existent awarenessbal space for its own spatial, territorial, geographic or geometric placement and existent. There is no escape for physical matter from this predicament or fate, regardless of its size, dimension, measurement or proportion.

Therefore, physical matter - including the physical matter of infinitesimally small sized singularity of Big Bang theory fame - can never be the source, creator, begetter or progenitor of current non-physical space or better still, of current non-physical cosmic space, not to mention, current non-physical human awareness, notwithstanding the theoretical claim of material scientists of the present-day.

Hence, only a non-physical, bodiless, dimensionless awareness which has no need of a pre-present or pre-existent cosmic space for its own spatial, geographic, geometric or territorial placement and existence, can be the source, creator, begetter or progenitor of current non-physical space or, more to the point, of current non-physical cosmic space and of the current non-physical awareness of human beings.

And this non-physical, bodiless, dimensionless awareness is none other than **the only one of its kind** or unique, timeless, bodiless, dimensionless awareness of infinite intelligence and emotion called **the creator** of current universe or **god** or whatever.

CONCLUSION

After analysis of both options namely, god on one hand and singularity of Big Bang theory fame on the other, with regards to the question as to which one amongst these two, is more likely to be the source of the current universe, one presents the following conclusion :-

The conclusion is that an incredible, absolutely non-physical, timeless, bodiless and dimensionless awareness of infinite intelligence and emotion called god or whatever is more likely to be the true source of the present-day universe rather than the absolutely physical and awareness, intelligence and emotion-devoid singularity of Big Bang theory fame, as theorised by the material scientists.

It will never be possible for material scientists (who are the author of this famous Big Bang theory) to physically get hold of and then directly experiment on the source of the current universe inside or, for that matter, outside or whatever, a large hadron collider and the like, irrespective of whether this source is the incredible, one hundred percent non-physical, timeless, bodiless and dimensionless awareness of infinite intelligence and emotion called god or whatever or, one hundred percent physical and awareness, intelligence and emotion-devoid singularity of Big Bang theory fame.

Material scientists and their physical-experiments plus their physical-cum-empiric proofs or better still, plus their physico-empiric-proofs   are not the 'be-all and end-all' or the 'alpha and omega' or the 'beginning and end' with regards to the mission or assignment of either locating or adjudicating the presence or absence of all the extant truths confronting mankind.

There are some extant truths such as human

awareness and god's awareness, not forgetting such emotions of human beings as love and hate, joy and sadness, aggression and depression etc. which confront mankind, the veracity of whose existence can only be established by awarenessbal-experiments as well as awarenessbal-cum-empiric-proofs or better still, by awarenessbal-empiric-proofs.

What has been said above can be put in another way.

The only one of its kind or unique, one hundred percent non-physical, timeless, bodiless and dimensionless awareness of infinite intelligence and emotion called **creator** of current universe or god or whatever is one such extant fact, the veracity of whose existence plus the veracity of whose innate nature can only be established beyond doubt by **awarenessbal-experiments** plus **awarenessbal-cum-empiric-proofs** or, better still, plus **awarenessbo-empiric-proofs** and not by physical-experiments and physical-cum-empiric-proofs or physico-empiric-proofs.

Material scientists can physically clasp and control plus scrutinise and scan as well as breakdown or splinter, that is to say, physically test or investigate or evaluate or experiment on only those items in the current universe which innately are one hundred percent physical in nature and which thus, innately also are one hundred percent, dimensional in nature.

Since all these items (items which are amenable to experiments by material scientists) are innately and absolutely physical in nature and therefore are also innately and absolutely dimensional in nature, it goes without saying that all these physical-cum-dimensional items or physico-dimensional items, therefore, are innately such things or items which perforce or by necessity, need or require a pre-existent space for their spatial placement and existence. That is to say, they all are such things which cannot exists in the absence of space or they all are such things for which the pre-presence of space is a sine qua non or an indispensable pre-condition if they are at all going to exist or occur in the universe. Pre-presence of space is a must for the existence of all such physico-dimensional things or beings.

Material scientists cannot physically clasp and control as well as breakdown or splinter with the help of such scientific gadgets as Large Hadron Collider and the like, any being which innately is one hundred percent non-physical in nature and thus is innately also one hundred percent non-dimensional or dimensionless in nature.

One such being in the universe which is innately and absolutely non-physical in nature and thus is innately and absolutely also non-dimensional or dimensionless in nature is the incredible, awareness of human beings.

With regards to this incredible human awareness, very few people of material science, if at all, want to talk about, except either to overlook it completely or to silently imply or indirectly hint, without explicitly committing or voicing that human awareness is one of the two effluents or excretions or, by-products or waste-products or, if it is preferred, is one of the two off-springs or progenies of the absolutely insentient plus dimensional, physical-matter-composed singularity of Big Bang theory fame.

As per material scientists, the other such effluent or excretion or, by-product or waste-product or, offspring or progeny of the absolutely insentient and dimensional, physical-matter-composed singularity of Big Bang theory fame, is none other than the incredible plus ubiquitous and absolutely non-physical cosmic space, without whose pre-presence or pre-existence, no physical-matter-composed entity, howsoever small in size, (and this includes the infinitesimally small sized physical singularity of Big Bang theory fame too), can have its spatial placement and existence in the current universe.

It has already been commented on above but it will be worthwhile repeating it here once again that, as per material scientists who have authored the Big Bang theory, the other entity in the current universe,

(apart from human awareness), which is also an effluent or excretion or, by-product or waste-product or, offspring or progeny of the one hundred percent, insentient plus dimensional, physical-matter-composed singularity of Big Bang theory fame, is the incredible plus ubiquitous and absolutely non-physical cosmic space; the incredible cosmic space which makes available to all absolutely insentient and dimensional plus absolutely physical-matter-composed entities, the vital or absolutely essential space or territory for their spatial or territorial placement and existence.

In total contrast to all absolutely physical-matter-composed entities of the current universe, (all of which, are innately and absolutely insentient in nature and hence they are all also innately and absolutely intelligence and emotion-devoid entities), the incredible human awareness of the current universe, is innately and absolutely, a non-physical plus intelligence and emotion-endowed entity.

Material scientists cannot physically clasp and control plus breakdown or splinter, such a non-physical and hence such a dimensionless entity as human awareness with the help of such physical-matter-composed scientific gadget as Large Hadron Colliders and the like, in the manner they are able to do with such physical and hence such dimensional, subatomic entities as protons and the like in order to unravel the secret of all the physical and hence of all the dimensional things of the current universe.

In addition to human awareness, there is another being which is also innately and absolutely non-physical in nature and hence, it is also innately and absolutely dimensionless in nature. This being is the incredible, one hundred percent non-physical, timeless, bodiless, dimensionless awareness of infinite intelligence and emotion called creator or god of the current universe or whatever.

As said above, this creator of the current universe called god or whatever, is innately and absolutely, non-physical in nature. Additionally, it is also, the only one of its kind or unique bodiless awareness anywhere. Thus, this creator of the current universe, called god or whatever, is innately and absolutely, the only one of its kind or unique bodiless and dimensionless awareness. In total contrast, human awareness is innately and absolutely a common-'o'-garden embodied and dimensionless awareness.

As a consequence of being innately and absolutely, the only one of its kind or unique, bodiless and dimensionless awareness, the incredible creator of the current universe called god or whatever, can and in fact does exist outside the realm, domain, field, arena or territory of space. That is to say, without the need or requirement of a pre-present or pre-existent space or territory for its own spatial or territorial placement and existence.

This is so because the creator of current universe or god or whatever has no need or requirement of a pre-present or pre-existent space or territory for its own spatial or territorial placement and existence due to its unique nature which is that of an incredible bodiless and dimensionless awareness.

This unique, bodiless and dimensionless essence of god's awareness or this unique, disembodied and dimensionless essence of god's awareness stands in total contrast to embodied and dimensionless essence of human awareness. That is to say, this unique, bodiless and dimensionless nature of god's awareness stands in total contrast to the nature of human awareness which, even though is also a dimensionless awareness in the manner of god's awareness, it nevertheless is an embodied awareness in total contradistinction to god's disembodied awareness.

On account of being innately and absolutely an embodied awareness, human awareness perforce or of necessity, always needs or requires a pre-present or pre-existent space or cosmic space, not for its own spatial placement and existence but for the spatial placement and existence of the physical-cum-dimensional body or physico-dimensional body inside which it resides or abides or inside which it has made its home or residence.

The only one of its kind or unique, absolutely non-physical, timeless, bodiless and dimensionless awareness of infinite intelligence and emotion called creator of the current universe or god or whatever is the source or the author of all the sentient as well as insentient plus time-bound, embodied and dimensional beings and things of the current universe such as embodied human awareness' on one hand and one hundred percent physical matter of the current universe on the other.

It has already been stated above but it will be worthwhile repeating it once again here that the current material scientists are capable of dealing only with those entities of the current universe which are one hundred percent physical in nature and thus, inherently are one hundred percent dimensional in nature.

These one- hundred -percent physical and therefore one hundred percent dimensional things of the current universe or, these one hundred percent physico-dimensional things of the current universe, as a consequence of their dimensional nature, perforce or by necessity, are in need or requirement of a pre-present or pre-existent space or territory for their spatial or territorial placement and existence.

Another innate characteristic of all physical-cum-dimensional items or, better still, all physico-dimensional items of the current universe is that they all, without exception, are time-bound or mortal in nature. In other words, none of them is timeless or immortal in nature in the manner of the non-physical-cum-dimensionless awareness, irrespective of whether the non-physical-cum-dimensionless awareness in question is the only one of its kind or unique, bodiless awareness of the creator of current universe or, the embodied awareness of a human being.

To recap.

Material science of any kind, including cosmology, astronomy, physics, chemistry, biology, astro-physics, astro-chemistry, astro-biology and the like, is not a tool which is competent to deal with such an incredible plus intangible or immaterial or insubstantial being as awareness; awareness which is innately and absolutely non-physical in essence or quintessence and as a result, which is also one hundred percent non-dimensional or dimensionless in essence or quintessence. It is of no consequence whether this awareness in question is the only one of its kind or unique, bodiless or disembodied awareness of the creator of current universe called god or whatever or, is the common-'o'-garden, embodied awareness of a human being.

What has been said above can be put in another way.

Material science is a tool which can clasp and control plus breakdown or splinter, if need be, only the physical and therefore only the dimensional component of a human being namely his physico-dimensional body or, if one prefers, only his physical or material body which is dimensional in configuration or silhouette.

In absolute contrast, material science of any kind cannot clasp or catch and constrain or control plus breakdown or splinter, the embodied but incredible, one hundred percent non-physical and therefore one hundred percent non-dimensional or dimensionless component of the human being namely his awareness, let alone the only one of its kind or unique, bodiless or disembodied plus non-physical and non-dimensional or dimensionless awareness of the creator of current universe called god or whatever.

Therefore, it is much easier or, better still, it is much more convenient for material scientists to see the current universe with the "colored glasses" of physical matter only. That is to say, through the prism of physicality or materiality only and not through the prism of aware-naiety or conscious-naiety which is the true essence or quintessence of the current universe.

Material sciences of all kinds are a part and parcel of

a dimensional and time-bound physical universe. Hence, along with the dimensional and time-bound, physical universe, material sciences too, one day, after a finite time, will dissolve back into the incredible creator of the current spacetime and physical universe called god or whatever.

To reiterate.

Material science, being a part and parcel of dimensional and time-bound physical universe, will never be able to physically hold in its test tubes or Large Hadron Colliders or whatever and then experiment on the absolutely non-physical, timeless, bodiless, dimensionless awareness of infinite intelligence and emotion called the creator of current universe or god or whatever. This is so, because this creator of current universe, in addition to being timeless, is also one hundred percent non-physical and non-dimensional or dimensionless in essence or quintessence in the manner of the incredible awareness of human beings.

It goes without saying that the being or the existence of this unique, timeless, bodiless and dimensionless awareness of infinite intelligence and emotion called creator of current universe or god or whatever, will predate, antedate, precede or antecede the being or the existence of its physical creation namely the dimensional and time-bound, physical or material or substantial universe.

May one remind those, who have reposed their faith absolutely in the current Big Bang theory of material scientists which theorises that its absolutely physical and yet incredibly and paradoxically non-spatial, non-territorial, non-geographic or non-geometric i.e. non-dimensional or dimensionless or beyond spacetime **singularity** of infinitesimally small size, of infinitely high temperature and of infinitely high density is the source of the current non-physical cosmic space, the current non-physical human awareness and the current physical matter in the present-day universe that this absolutely physical and yet incredibly and paradoxically, non-spatial, non-territorial, non-geographic or non-geometric i.e. non-dimensional or dimensionless or beyond spacetime **singularity** of Big Bang theory fame no longer exists in its original form namely that of an absolutely physical and yet incredibly and paradoxically non-spatial, non-territorial, non-geographic or non-geometric i.e. non-dimensional or dimensionless or beyond spacetime form. Instead, it now exists as the current non-physical cosmic space, the current non-physical human awareness and the current physical matter of the present-day universe.

As per Big Bang theory of the material scientists, absolutely physical and yet incredibly and paradoxically non-spatial, non-territorial, non-geographic or non-geometric i.e. non-dimensional or dimensionless or beyond spacetime **singularity** of infinitesimally small size, of infinitely high temperature and of infinitely high density has transformed itself into the present-day universe following the event of Big Bang and the process of inflation plus the activities of such laws of Mother nature as gravity, anti-gravity, quantum mechanics and the like on one hand and Darwinian law of evolution on the other. Therefore, at the present moment, this absolutely physical and yet incredibly and paradoxically non-spatial, non-territorial, non-geographic or non-geometric i.e. non-dimensional or dimensionless or beyond spacetime **singularity** of Big Bang theory fame is unavailable to material scientists in its original form for direct or first- hand experimentation such as its physical cleavage, splitting or splintering inside such scientific gadgets as Large Hadron Colliders and the like. This absolutely physical cum theoretical **singularity** of Big Bang theory fame is now available to them i.e. to material scientists for such physical experiments only as the physical matter of the present-day universe.

Material scientists totally overlook, disregard, ignore, pay no heed to or turn a blind eye to the other two ingredients of the present-day universe namely the non-physical cosmic space and the non-physical human awareness.

They i.e. the material scientists show keenness, avidness, ardour or fervour only for the physical matter of the present-day universe because the latter namely the physical matter of the present-day universe allows them to do physical experiments on it in such scientific gadgets as The Large Hadron Collider and the like and also allows them to construct their such famous physical theories about the current physical universe and its physical matter as Big Bang theory and the like. There is no such scope, opening, opportunity, possibility, leeway or latitude for the present-day material scientists with respect to such non-physical constituents of the present-day universe as cosmic space and human awareness. Therefore, it comes as no surprise that the present-day material scientists prefer to overlook or turn a blind eye to the latter two ingredients of the current universe namely the non-physical cosmic space and the non-physical human awareness.

Now let's talk about the incredible, timeless, bodiless and dimensionless awareness of infinite intelligence and emotion called the creator of current universe or god or whatever.

As an alternative to the Big Bang theory, if one has reposed one's faith absolutely into the idea of an incredible, timeless, bodiless and dimensionless awareness of infinite intelligence and emotion called god, creator or whatever as being the source of the current universe, then even in this scenario, the material scientists cannot directly or first-hand approach and apprehend this incredible, timeless, bodiless and dimensionless awareness of infinite intelligence and emotion called god, creator or whatever in order to directly, touch or feel such a god or creator and then deconstruct or dissect or, cleavage or splinter such a god or creator so as to investigate, analyse and define its innate attributes inside such a physical scientific gadget such as The Large Hadron Collider and the like.

This is so because physical and dimensional things such as The Large Hadron Collider and the like can never physically reach out to such a non-physical and dimensionless being or thing as the source of the current universe (whatever that may be) for the purpose of direct physical inspection, investigation and analysis.

Such inspection, investigations and analyses of this incredible source of the current universe will therefore, perforce or of necessity, always be, indirect only. That is to say, by deduction or, by extrapolation or, of provisional nature or, of probationary nature or, of trial-and-error nature or, of interpretational nature or, of hypothetical nature or, of experimental nature or, of empirical nature or, utmost, of the nature of a scientific theory such as Big Bang theory and the like.

Thus, either way, the material scientists and their material science can and will only wonder and surmise indirectly, in the way all humanity does, as to who or what this ultimate source of the present-day spacetime and physical cum awarenessbal universe really is.

This is what material scientists actually do when they are asked to define the attributes of their physical and yet incredibly and paradoxically non-spatial, non-territorial, non-geographic or non-geometric i.e. non-dimensional or dimensionless or beyond spacetime **singularity** which they regard as being the ultimate source of the current spacetime and physical cum awarenessbal universe in contradistinction to the incredible, one hundred percent, non-physical, timeless, bodiless and dimensionless awareness of infinite intelligence and emotion called creator, god or whatever.

Material scientists theorise that this **singularity** of Big Bang theory fame existed per se or as such 13.7 billion light years ago only.

At that instant i.e. 13.7 billion light years ago, this theoretical **singularity** of Big Bang theory fame was an infinitely hot, infinitely dense and infinitesimally small physical thing. It then, despite being absolutely physical in nature, incredibly and paradoxically existed without the need or requirement of a pre-present or pre-existent cosmic space.

In other words, **singularity** of Big Bang theory fame, 13.7 billion light years ago, incredibly and paradoxically existed without the need or requirement of a pre-present or pre-existent cosmic space for its own spatial placement and existence despite being innately and absolutely physical in nature or, in spite of being innately and absolutely physical in nature.

What has been said above vis-a-vis **singularity** of Big Bang theory fame can be put another way.

Material scientists theorise that **singularity** of Big Bang theory fame was infinitely hot, infinitely dense and infinitesimally small physical thing which existed as such or, alone or, by itself or, per se in this original form 13.7 billion light years ago without the presence, existence or being of the present-day cosmic space because this incredible physical **singularity's** presence, existence or being, predated the presence, existence or being of the present-day cosmic space.

Material scientists further theorise that through the event of Big Bang and the process of inflation or expansion of the physical substance of singularity (in the manner of inflation or expansion of a balloon) plus through the gradual cooling of the inflated or expanded physical substance of singularity and the forces of such laws of Mother Nature as the law of gravity, the law of anti-gravity, the law of quantum mechanics and the like on one hand and the Darwinian law of evolution on the other, this physical and yet incredibly and paradoxically non-spatial, non-territorial, non-geographic or non-geometric i.e. non-dimensional or dimensionless or beyond spacetime **singularity** of infinitesimally small size, of infinitely high temperature and of infinitely high density, transmuted, transformed or metamorphosed itself into the present-day non-physical cosmic space and the present-day non-physical human awareness of current universe on one hand and the present-day physical matter of current universe on the other.

By the way, may one point out, in case one has failed to notice, that the use of the phrase **infinitesimally small** on the part of material scientists to describe one of the three cardinal attributes of their physical **singularity** of Big Bang theory fame, amounts to the use on their part of an **euphemism** or **understatement** or if it is preferred, amounts to the use on their part of a **mild** or **indirect expression** in order to signify, imply or symbolise the characteristic of **dimensionless-ness, non-spatial-ness, non-territorial-ness, non-geographical-ness, non-geometrical-ness** or **beyond spacetime-ness** vis-a-vis or with regards to their physical **singularity** without explicitly saying so.

Thus, material scientists are implying, without explicitly saying so, that their physical **singularity** of Big Bang theory fame is as **dimensionless** in nature as the absolutely non-physical, timeless, bodiless and dimensionless awareness of god is as well as the absolutely non-physical, timeless but embodied, dimensionless awareness of man is.

To reiterate.

Material scientist's theory is that their **absolutely physical,** infinitesimally small, infinitely hot and infinitely dense **singularity** of Big Bang theory fame is as **dimensionless** in nature as the **absolutely non-physical** awareness of god is as well as the **absolutely non-physical** awareness of human being is.

In other words, material scientists theorise that by virtue of being **infinitesimally small sized,** the **absolutely physical** singularity of Big Bang theory fame becomes entitled to be endowed with one of the **key characteristics** of **absolutely non-physical** awareness of god as well as the **absolutely non-physical** awareness of human beings namely the characteristic of being **dimensionless** despite being **absolutely physical** in nature.

As a result, as per material scientists, the infinitesimally small sized but innately and absolutely physical-cum-dimensional or physico-dimensional singularity of Big Bang theory fame also becomes entitled to be in existence in reality without the need or requirement of pre-present or pre-existent space or, more to the point, pre-present or pre-existent cosmic space for its spatial placement and existence in the manner of innately and absolutely 'non-physical-cum-non-dimensional' or 'non-physical-cum-dimensionless' awareness of god, despite the fact that this singularity of Big Bang theory fame is theorised by the material scientists as being innately and absolutely physical and therefore, inevitably or unavoidably dimensional in nature.

What has been said above vis-a-vis singularity of Big Bang theory fame can be put in another way.

As a result, according to material scientists, the infinitesimally small sized but innately and absolutely 'physical-cum-dimensional' or 'physico-dimensional' singularity of Big Bang theory fame also becomes entitled of being accepted by mankind as an absolutely real thing which can categorically, legitimately, rightfully or unquestionably exist without the need or requirement of the pre-present or pre-existent space or, more to the point, the pre-present or pre-existent cosmic space for its own spatial placement and existence in the manner of the innately and absolutely 'non-physical-cum-non-dimensional' or 'non-physical-cum-dimensionless' awareness of god, despite the fact that this singularity of Big Bang theory fame is theorised by the material scientists as being innately and absolutely physical and therefore, inevitably or unavoidably dimensional in nature.

Furthermore, as per material scientists, this innately and absolutely 'physical-cum-dimensional' or 'physico-dimensional' singularity of Big Bang theory fame thus, also becomes entitled to be accorded the credit of being the ultimate source, not only of the present-day physical matter in the current universe but also of the present-day absolutely non-physical cosmic space, as well as the present-day absolutely non-physical human awareness in the current universe.

However, by all criteria or yardstick of today's universe, the above expectation of material scientists with regards to their theoretical singularity of Big Bang theory fame is absolutely illogical or unreasonable because such an expectation on their part does not at all accord or gel with the present-day reality vis-a-vis anything which is one hundred percent physical in make-up. All such physical things, irrespective of their size, will inevitably be deemed to be one hundred percent dimensional in configuration. Therefore, all such physical things, (irrespective of their size), will be judged to be in need of a pre-existent space or cosmic space for their own spatial placement and existence. Hence, it is impossible for such a physical thing as **singularity** of Big Bang theory fame, (not withstanding its infinitesimally small size), to be the source or creator of such a thing as cosmic space.

To recap.

Singularity of Big Bang theory fame is theorised by the material scientists as being one hundred percent physical in nature. Therefore, such a physical singularity will inevitably be one hundred percent dimensional in configuration as per the current norm of today's universe. Therefore, such a 'physical-cum-dimensional' or 'physico-dimensional' entity as singularity of Big Bang theory fame **cannot be real** if it is theorised by the material scientists to exist in the absence of present-day cosmic space because a physical thing like singularity, notwithstanding its infinitesimally small size, will always need or require a pre-existent cosmic space for its own spatial placement and existence.

Therefore, once again one repeats here that such a physical thing as singularity **cannot be real** if is theorised by the material scientists to exist in the absence of present-day cosmic space.

Hence, the Big Bang theory which theorises that such a physical thing as singularity, notwithstanding its infinitesimally small size, is the source or creator of the present-day cosmic space is not credible. This is because physical singularity itself will need cosmic space for its own spatial placement and existence.

Material scientists will therefore have to find some

other explanation for such physical phenomena as cosmic microwave background radiation (CMBR), cosmological redshift i.e. expanding universe & predominance of lighter elements such as hydrogen and helium in the present-day observable universe.

EPILOGUE

Irrespective of what or who is the true source of the troika called cosmic space, human awareness and physical matter in the current universe, the vital point to take note of is that this source of cosmic space, human awareness and physical matter in the current universe will definitely possess the following characteristics :-

1. It will be timeless.

2. It will be dimensionless.

3. It will not personally, directly, first hand or face-to-face be available to material scientists for such physical-cum-scientific experiments as cleavage, splitting, splintering, smashing or shattering in the way such subatomic particles as protons of the current physical realm are subjected to by the present-day material scientists in such scientific gadgets as Large Hadron Collider and the like.

4. Its existence will pre-date the existence of the current troika called cosmic space, human awareness and physical matter.

5. It will therefore be such an incredible being or thing that it will be in a position to exist without the need of the present-day cosmic space or, in the total absence of the present-day cosmic space. Or better still, it will be such an incredible being or thing that it will be capable of existing in an absolutely cosmic-space-devoid-domain.

It has been mentioned above that this incredible source of the troika called cosmic space, human awareness and physical matter of the current universe will be dimensionless in configuration.

As a result of being dimensionless in configuration, this incredible source of cosmic space, human awareness and physical matter will therefore be innately and absolutely blessed, endowed or gifted or, better still, accredited, enabled or empowered to exist in the total absence of cosmic space of which it is the source, creator, maker or progenitor.

What has been said above can be put in another way.

Since this source of the troika called cosmic space, human awareness and physical matter in the current universe will intrinsically be one hundred percent dimensionless in configuration, it will be absolutely competent to exist or it will be absolutely cut out to exist without the need of a pre-present or pre-existent cosmic space for its own spatial placement and existence; cosmic space of which it is the source, creator, maker or progenitor.

Hence, irrespective of whether one believes and accepts that the source of the troika called cosmic space, human awareness and physical matter is the one hundred percent non-physical, timeless, bodiless and dimensionless awareness of infinite intelligence and emotion called creator, god or whatever or it is the one hundred percent physical, infinitesimally small, infinitely hot and infinitely dense singularity of Big Bang theory fame, the inescapable truth is that this source or creator's own existence, first and foremost or above all, will pre-date the existence of the current cosmic space. As a result, the intrinsic nature of this incredible source or creator of the current cosmic space, not forgetting, the current human awareness and current physical matter, will need to be or, will have to be, of such a unique kind that it will be one hundred percent non-physical and non-dimensional or dimensionless in configuration.

As a consequence of innately being one hundred percent non-physical and non-dimensional or dimensionless in nature, this source or creator of the current cosmic space, the current human awareness and the current physical matter will become empowered to exist in the total absence of cosmic space. It will therefore, also become fully qualified to be accepted as the true source or the creator of the current cosmic space, the current human awareness and the current physical matter.

The implication of all that has been said above is that, irrespective of whether the author of the present-day cosmic space is the one hundred percent non-physical, timeless, bodiless and dimensionless awareness of infinite intelligence and emotion called creator, god or whatever or, the one hundred percent physical, infinitesimally small, infinitely hot and infinitely dense singularity of Big Bang theory fame, it perforce or of necessity, will have to be of such a unique nature that it will not have a need or requirement of a pre-present or pre-existent space or, more to the point, cosmic space for its own spatial placement and existence and thus, it will be totally capable of existing in the absolute absence of cosmic space.

Therefore, one can logically conclude that the following features must attend or accompany **singularity** of Big Bang theory fame before it can be considered as being the true source of the current cosmic space, the current human awareness and the current physical matter, despite the big hope and expectation of Big Bang theory of the material scientists that this one hundred percent physical **singularity** is the true source of the current cosmic space, of the current human awareness and the current physical matter and not the incredible, one hundred percent non-physical, timeless, bodiless and dimensionless awareness of infinite intelligence and emotion called creator, god or whatever.

The features which must attend or accompany **singularity** of Big Bang theory fame before it can be considered as being the true source of the current cosmic space, the current human awareness' and the current physical matter of the present-day cosmos are the following :-

1. Firstly, the existence of one hundred percent physical, infinitesimally small, infinitely hot and infinitely dense singularity of Big Bang theory fame will have to pre-date the existence of the current cosmic space.

2. Secondly, physical singularity of Big Bang theory fame will have to have such an extraordinary innate nature that despite being one hundred percent physical in composition, it will enigmatically or paradoxically be dimensionless in configuration and thus will be capable of existing in the total absence of cosmic space meaning thereby it will have no need of a pre-present or pre-existent cosmic space for its own spatial placement and existence which is quite unlike the way in which all physical things of the present-day universe behave, irrespective of their size or, better still, irrespective of whether they are of an infinitesimally small size (as the **physical singularity** of Big Bang theory fame has been theoretically

conceived by the material scientists) or are of an immensely big size.

What has been said above can be put in another way.

Each and every physical thing in the present-day universe is dimensional in configuration. None of them is dimensionless.

That is to say, as per immutable law of Mother Nature at the present moment, all things of the present-day universe which are physical in nature (irrespective of whether they are of extremely minuscule size i.e. sub-atomic particles like electrons, protons, neutrons and the like or, of extremely massive size i.e. astral objects like planets, stars, galaxies and the like are dimensional in configuration. There is no exception to this unyielding rule.

In other words, as per immutable law of Mother Nature at the present moment, anything which is one hundred percent physical in composition can never ever be dimensionless in configuration.

Let one repeat the above statement once again. "As per immutable law of Mother Nature at the present moment, anything which is one hundred percent physical in composition can never be dimensionless in configuration".

Thus, all things physical, without exception, will require or need the pre-presence or pre-existence of cosmic space for their spatial placement and existence because their existence will be impossible without the pre-presence or pre-existence of cosmic space. In other words, anything which is physical in composition, including **singularity** of Big Bang theory fame can never be the source of cosmic space. That's final. Full stop.

Hence, Big Bang theory of material scientists which theorises, without saying so in so many words, that its heart namely, the infinitesimally small, infinitely hot and infinitely dense plus one hundred percent **physical singularity** is capable of existing without the need or requirement of a pre-present or pre-existent cosmic space, seems totally out of sync with the present-day reality or seems totally incompatible with the present-day reality which is that no entity which is one hundred percent physical in composition, (irrespective of its size or, better still, irrespective of its infinitesimally small size) can never ever exist in the absence of cosmic space or, without the pre-presence or pre-existence of cosmic space.

Therefore, the above theoretical conclusion or, better still, the above theoretical hope of the material scientists who have given birth to Big Bang theory that the heart of the Big Bang theory namely, the one hundred percent **physical singularity** of infinitesimally small size will be capable of existing in the total absence of cosmic space or, will be capable of existing without the need or requirement of pre-present or pre-existent cosmic space, is extremely perplexing, puzzling or confounding, not to mention counterintuitive, illogical, nonsensical, untenable or absurd. As a matter of fact, this is absolutely impossible because it is against the current immutable law of Mother Nature which is obvious to all and it is that all things physical, irrespective of their size, will always be dimensional in configuration. And therefore, they all will be incapable of existing in the total absence of cosmic space or, they all will always need or require a pre-present or pre-existent cosmic space for their own spatial placement and existence.

Unfortunately, material scientists do not come out in the open to explain the above-mentioned paradox, puzzle, enigma or conundrum present in their Big Bang theory.

That is to say, unfortunately material scientists do not come out in the open to explain the above mentioned paradox with regards to their infinitesimally small, infinitely hot and infinitely dense theoretical **singularity** namely that how this one hundred percent physical cum theoretical **singularity** of Big Bang theory fame (notwithstanding its infinitesimally small size), can be dimensionless in configuration so that it can exist in the absolute absence of cosmic space or without the need or requirement of a pre-present or pre-existent cosmic space.

What has been said above can be put in another way.

Material scientists do not answer transparently, the most vital question, "how such a physical cum theoretical thing like **singularity** of Big Bang theory fame - even if it is theorised by the material scientists to be of infinitesimally small size - can be dimensionless in configuration?"

Existence of an entity in the absolute absence of cosmic space is only possible if that entity is innately one hundred percent dimensionless in configuration. There is no exception to this rule.

The only one of its kind or unique, innately and absolutely non-physical, timeless, bodiless and dimensionless awareness of infinite intelligence and emotion called **the creator** of current cosmic space, current human awareness' and current physical matter is the only entity anywhere which is innately and absolutely dimensionless in configuration and therefore is the only entity which is capable of existing in the absolute absence of cosmic space. It therefore, also is the only entity which can be and in fact, is the creator of current cosmic space and no one else including the absolutely physical cum theoretical **singularity** of Big Bang theory fame.

To sum up.

It is not possible for anything which is physical in nature to be dimensionless in configuration even if it is of infinitesimally small size.

The vital question, "how such a physical cum theoretical thing as **singularity** of Big Bang theory fame, (notwithstanding its infinitesimally small size), can ever exist in the absolute absence of cosmic space", has remained unanswered by the material scientists for far too long.

Therefore, it will not be improper for people to question the material scientists about their view that the heart of their Big Bang theory namely, the infinitely hot, infinitely dense and infinitesimally small, one hundred percent physical cum theoretical **singularity** is the ultimate source of the present-day troika namely, the cosmic space, human awareness' and physical matter and not **the only one of its kind** or unique, one hundred percent non-physical, timeless, bodiless, and dimensionless awareness of infinite intelligence and emotion called god or whatever.

Now one hopes that all people including the material scientists will be able to see clearly the presence of the above mentioned paradox in the

Big Bang theory that an infinitely hot, infinitely dense and infinitesimally small, one hundred percent physical cum theoretical thing called **singularity** is the source of the present-day troika namely the cosmic space, human awareness' and physical matter and not **the only one of its kind** or unique, innately and absolutely non-physical, timeless, bodiless and dimensionless awareness of infinite intelligence and emotion called god or whatever.

~*~*~*~*~

ADWAIT-VEDANTA &CONSCIOUSNESS-1

Consciousness is the only entity inside the current or contemporary cosmos which is ageless, deathless, timeless, immortal or eternal, irrespective of whether it is the **human consciousness** extant or present in a **human-physical-body** or, it is **god's consciousness** extant or present in the form of the current **cosmic space.**

Here one must point out that the word **'god'** has been used above in order to designate or symbolise the **consciousness** of the **creator, maker** or **progenitor** of the **current** or the **contemporary cosmos.** However, human beings are free to call the **consciousness** of the creator of the current cosmos by any name they fancy. The name by which the **consciousness** of the creator of the current cosmos is called by human beings is immaterial. What is truly or vitally important for human beings to understand is that the **consciousness** is the only entity inside the current cosmos which is ageless, deathless, timeless, immortal or eternal, irrespective of whether it is the **human consciousness** extant or present in a **human-physical-body** or, it is **god's consciousness** extant or present in the form of the current **cosmic space.**

The **consciousness** of god or creator of the current cosmos is extant or present in the current cosmos in the form, manifestation or incarnation of the ubiquitous and the infinite *field of consciousness,* called **cosmic space.**

It is only the physical matter extant or present in the current cosmos which is evanescent, ephemeral, transient or time-bound and not the **consciousness.**

To recap.

The ageless, deathless, timeless, immortal or eternal **consciousness** extant or present in the current cosmos is of two kinds :-

Firstly, it is extant or present in the current cosmos in the form of **embodied consciousnesses** of such **conscious beings** of the current cosmos as human beings.

Secondly, it is extant or present in the current cosmos in the **non-embodied, disembodied, disbodied** or **bodiless** form as the ubiquitous and the infinite *field of consciousness* called **cosmic space** which is the expanded, distended, dilated or inflated form, version, manifestation or incarnation of the **original, bodiless** and **dimensionless consciousness** of creator, god or whatever of the current cosmos.

Each and every moment of its existence in the current cosmos, every **human consciousness** must, without fail, remember, recall or think of the fact of its ageless-ness, deathless-ness, timelessness, immortality, imperishability or indestructibility if it wants to reduce its fear in the future at the moment of the inevitable death of its physical body.

That is to say, every human consciousness must always remember that it itself will never die even though the physical body inside which it presently lives, resides, abides or dwells will one day surely die.

Human consciousness must also understand and internalise the reason why it will never die or why it is immortal or eternal in the manner of its source, fountainhead or **point of supply** namely the creator of the current cosmos or god or whatever.

The reason is that, in contrast to the physical body inside which human **consciousness** presently abides or resides, it itself is an **absolutely pristine segment** of the ubiquitous and the infinite *field of consciousness* called **cosmic space.** By being an **absolutely pristine segment** of the ubiquitous and the infinite *field of consciousness* called **cosmic space,** it is as eternal, immortal, imperishable, indestructible, timeless or deathless as **cosmic space.**

Cosmic space, in its turn, is the expanded, distended, dilated or inflated form, version, manifestation or incarnation of the immortal, **dimensionless consciousness** of the creator of the current cosmos.

It is the **consciousness** and **consciousness** alone and not the physical matter of the current cosmos which is innately capable of existing or inhering and in fact it does exist or inhere timelessly, endlessly, eternally or perpetually. Physical matter (and all things made of physical matter) in the current cosmos, on the other hand, is innately time-bound, transient, temporary, evanescent or ephemeral. Physical matter of the current cosmos is innately time-bound, transient, temporary, evanescent or ephemeral because it is made of a **condensed, compressed, or compacted** *section or segment* of the ubiquitous and the infinite **field of consciousness** called *cosmic space* which in turn, is the **expanded, distended, dilated or inflated** form, version, manifestation or incarnation of the immortal or eternal **consciousness** of the creator of the current cosmos or god or whatever.

Human **consciousnesses** of the current cosmos must also realise that the entity called **consciousness** will also exist, inhere or will be extant or present when the current cosmos will no longer exist, inhere or will no longer be extant or present in the future.

In other words, the entity called **consciousness** will exist, inhere or will be extant or present in the absence of the current cosmos or when the current cosmos will disappear in the future.

The disappearance of the current cosmos must occur or rather will definitely occur sometime in the future as per mood, whim or fancy of its creator.

The disappearance of the current cosmos is inevitable in future because it began 13.7 billion light years ago as per its creator's will. Everything which has a beginning, will also have an end at some point in future. This is the rule of the creator of the current cosmos.

It has already been said earlier but it will be worthwhile repeating the same once again that **consciousness** in the current cosmos exists, inheres or is extant or present in the form of **consciousness** of such *embodied, conscious beings* as human beings on one hand and in the form of *non-embodied*, ubiquitous and the infinite **field of consciousness** called ***cosmic space*** on the other. The ***non-***

embodied, ubiquitous and the infinite **field of consciousness** called *__cosmic space__* is the **expanded, distended, dilated or inflated** form, version, manifestation or incarnation of the **original, dimensionless consciousness** of the creator of the current cosmos.

Both inside the current cosmos and outside or beyond the current cosmos, **consciousness** is the only **self- aware** truth.

To reiterate.

Consciousness is the only reality anywhere which is innately and absolutely **self-aware.** Physical matter, extant in the current universe, on the other hand, is innately and absolutely not **self -aware.**

In other words, **consciousness** is the only presence anywhere which is **aware** of its own existence as an **aware being.** It is also **aware** of other beings which themselves are also **self- aware.**

Moreover, **consciousness** is also the only entity anywhere which is **aware** of the existence of those truths which are themselves not innately **self -aware** or which are themselves not innately **aware** of their own being, for example, the insentient physical matter of the current cosmos.

To sum up.

Consciousness is the only truth anywhere which has the inherent or the intrinsic ability to say to itself:- "I exist, I exist, I exist". It also has the inherent or the intrinsic ability to say to other truths:-"You exist, You exist, You exist" or "it exists", "it exists", "it exists".

Anything which is innately and absolutely not **self-aware** is designated by the **self- aware** consciousness as **physical matter.**

Thus the current cosmos is made up of two elementary ingredients namely:-

Innately and absolutely **self- aware** consciousness.

Innately and absolutely not **self- aware** physical matter.

Innately and absolutely **self- aware** consciousness of the current cosmos is of two kinds namely:-

Embodied, self-aware consciousness. For example, consciousness of such **embodied** conscious beings of the current cosmos as human beings.

Non-embodied, disembodied or bodiless, self-aware consciousness i.e. the ubiquitous and the infinite **field of consciousness** called *cosmic space* which is the **expanded, distended, dilated or inflated** form, version, manifestation or incarnation of the **original, dimensionless consciousness** of the creator of the current cosmos.

~*~*~*~*~

ADWAIT–VEDANTA & CONSCIOUSNESS - 2

Self- awareness is **alpha and omega** or **supreme** feature which separates the sentient **consciousness** from the insentient **physical matter** of the current cosmos.

Even before the birth or nascence of the current cosmos, there always was, always is and always will be an **existence** because there never was, never is and never will be the state of **absolute non-existence** or **absolute non-being** or **absolute nothingness** or, if one prefers, there never was, there never is and there never will be the state of **absolute cipher, zero** or

shoonya.

The **eternal existence,** about which one has touched upon above or hinted above, abides, endures, occurs or obtains in the form of an **eternal consciousness** and not in the form of any kind of physical matter.

Thus, there always was, always is and always will be an **eternal existence** and this **eternal existence** is an **eternal consciousness** of an incredible or awe-inspiring nature.

Physical matter, on the other hand, is not an eternal existence.

That is to say, physical matter is merely a transient, temporary or short-lived existence because it sometimes exist and sometimes does not. Whenever it exists, it exists merely as a **daydream, reverie, imagery, dreamry** or **fantasy** of the above - mentioned **eternal consciousness,** nothing more nothing less.

To repeat.

Physical matter, irrespective of its size, shape, style or idiosyncrasy, will always be nothing but a transient, temporary or short-lived existence and not an ageless, deathless, timeless or eternal existence in the manner of the **eternal consciousness.**

As said earlier, the **eternal existence** always was, always is and always will be the **eternal consciousness** only and nothing else.

Now let's explore this **eternal existence** aka **eternal consciousness** in more detail.

There are two versions or manifestations of this **eternal existence** aka **eternal consciousness.** These are described below.

This incredible **eternal existence** aka **eternal consciousness** sometimes exists **all alone** i.e. **all by itself,** without containing any kind of cosmos or universe inside itself as its **daydream, reverie, imagery, dreamry** or **fantasy.**

At other times this incredible **eternal existence** aka **eternal consciousness** exists while at the same time accommodating or containing inside itself a cosmos or universe as its **daydream, reverie, imagery, dreamry** or **fantasy.** This is the state of affairs with this **eternal existence** aka **eternal consciousness** at the present moment in time.

All **human consciousnesses** must always be cognisant of the fact that whenever they perceive and experience a **cosmos** or **universe** in their wakeful state, as is the case at the present moment, the existence of such a **cosmos** or **universe** is always inside an incredible or awe-inspiring **eternal existence** aka **eternal consciousness** only and nowhere else.

That is to say, whenever **human consciousnesses** perceive and experience a **cosmos** or **universe** in their wakeful state, as is the case at the present moment, they must always be cognisant of the fact that this **cosmos** or **universe** abides, exists, occurs or obtains inside an **eternal existence** aka **eternal consciousness** only and nowhere else.

Furthermore, all **human consciousnesses** must always also be aware of the fact that this **cosmos** or **universe**, which they perceive and experience in their wakeful state; as is the case at present; abides, exists, occurs or obtains inside this **eternal consciousness,** purely as a transient, temporary or short-lived **daydream, reverie, imagery, dreamry** or **fantasy** of this **eternal consciousness** and nothing else.

That is to say, this **cosmos** or **universe** does not abide, exist, occur or obtain as an ageless, deathless, timeless, eternal or immortal reality or truth in the manner of the **eternal consciousness** whose transient, temporary or short-lived **daydream, reverie, imagery, dreamry** or **fantasy** they are.

The current or contemporary **cosmos** or **universe;** which all human consciousnesses of the present-day perceive and experience all through their lives during their wakeful state; has been created or has been given birth or nascence by the incredible or awe-inspiring **eternal consciousness** in question inside **itself** through the instrumentality of **daydreaming, reverieing, imagerying** or **dreamrying** on its part, and nothing else.

ADWAIT-VEDANTA & CONSCIOUSNESS - 3

A. *Eternal consciousness existing all alone or all by itself.*

When one says, **eternal consciousness** existing **all alone** or **all by itself** one means that **eternal consciousness** is existing sans cosmos or existing without cosmos.

What has been said above can be put in another way.

When one says, **eternal consciousness** existing **all alone** or **all by itself,** one means that **eternal**

consciousness is existing without having, possessing or fostering the current or contemporary cosmos inside itself as its **daydream, reverie, imagery, dreamry** or fantasy.

The state of **eternal consciousness** sans cosmos or without cosmos or, the state of **eternal consciousness** existing **all alone** or **all by itself** is described as the incredible or awe-inspiring and **the only one of its kind** or unique, *timeless, bodiless and dimensionless* **consciousness, awareness, sentience** or **mind of infinite intelligence and emotion.**

B. *Eternal consciousness existing with its daydream-stuff-composed cosmos extant, present or in existence inside it.*

In this state, the **eternal consciousness** exists with its **cosmos** or **universe** extant or present inside it as its **daydream, reverie, imagery, dreamry or fantasy.** This **cosmos** or **universe** has been created or formed by this incredible **eternal consciousness** inside **itself** through the instrumentality of **daydreaming, reverieing, imagerying** or **dreamrying** on its part, nothing more nothing less. This is the state of affairs at the present moment in time.

~*~*~*~*~

ADWAIT-VEDANTA & CONSCIOUSNESS - 4

Human consciousness must understand clearly that the **eternal existence** never was, never is and never will be the **insentient, unconscious** or **unaware** physical matter of the current cosmos, **physical matter** which is incapable of any kind of feeling and understanding and which is dead to its own existence as well as dead to the existence of everything else around it.

Instead, the **eternal existence,** always was, always is, and always will be the unique or **the only one of its kind,** *timeless, bodiless and dimensionless* **consciousness, awareness, sentience or mind of**

infinite intelligence and emotion called creator, god or whatever of the current cosmos.

What has been said above can be put in another way.

The **eternal existence** always was, always is and always will be **the only one of its kind** or unique, innately and absolutely **self- aware, *non-physical, timeless, bodiless and dimensionless* consciousness, sentience** or **mind** of infinite intelligence and emotion called creator, god or whatever, which at the present moment is extant, present or in existence in its metamorphosed or transmuted form called the current or contemporary cosmos.

The current cosmos consists of three basic ingredients which are as follows.

1) **The current cosmic space.**

2) **The current physical matter.**

And last but not least :-

3) **The current assortment or medley of a large number of individual, separate, or discrete, embodied consciousnesses such as human consciousnesses etc.**

Before it metamorphosed or transmuted itself into its

present form namely the current cosmos, the incredible **eternal existence** namely the creator, god or whatever of the current cosmos was extant or present in its *original, primordial, native* or *first* form.

This *original, primordial, native* or *first* form or version of the **eternal existence** namely the creator, god or whatever of the current cosmos, is described as **the only one of its kind,** innately and absolutely **self-aware, non-physical, timeless, bodiless and dimensionless consciousness, sentience** or **mind** of infinite intelligence and emotion.

The above described **original state** of creator, god or whatever of the current cosmos is also designated as **supra-transcendental state** of the creator, god or whatever of the current cosmos.

In other words, before this incredible creator, god or whatever metamorphosed itself into the current or contemporary cosmos, it existed in its **original, native, primordial or first** form or state which is designated as its **supra-transcendental** form or state.

To restate.

The original form or state of creator, god or whatever of the current cosmos i.e. before it metamorphosed itself into the current cosmos, is labelled as its **supra-transcendental form** or **state.**

Therefore, it can be said that the expression **eternal**

existence in the **supra-transcendental realm** or **domain** signifies or stands for **eternal consciousness, awareness, sentience** or **mind** only and nothing else.

In the **supra-transcendental** *realm* or *domain* there is no access or entry to physical matter, whatever be its size, shape or characteristic. Physical matter, irrespective of its size, shape or characteristic, will always be an **infra-transcendental** truth meaning thereby, it will always be a transient, temporary or short-lived truth and never an eternal truth.

The role of the transient, temporary or short-lived **infra-transcendental** truth namely the physical matter is always, only to create or generate, transient, temporary or short-lived, variety, diversity or heterogeneity inside the **nondescript, featureless, monochrome** or **unvarying, eternal consciousness,** nothing more nothing less.

What has been said above can be put in another way.

The expression **eternal existence** in the **supra-transcendental realm** or **domain** never signifies or stands for physical matter because physical matter takes birth inside this **eternal existence** merely as its **daydream, reverie, imagery, dreamry** or **phantasy** and nothing else and thus it never is and never will be eternal. It is condemned to be transient, temporary or short-lived only forever and nothing else.

~*~*~*~*~

ADWAIT–VEDANTA & CONSCIOUSNESS - 5

What is designated as **physical matter** by **human consciousnesses** in the current cosmos is merely a **daydream-stuff-composed** reality, which has been given birth by the ubiquitous and the infinite **field of consciousness** called **cosmic space** through the activity of **daydreaming** on its part.

The entity labelled as **cosmic space** by **human consciousnesses** in the current cosmos, on the other hand, is the expanded, distended, dilated or inflated **form**, **version** or **state** of **bodiless** and **dimensionless consciousness** of the creator, god or whatever of the current cosmos, nothing more nothing less.

As regards the current cosmos, it essentially consists of the following three ingredients :-

1) Cosmic space.

2) A large assortment or medley of separate, discrete or individual, embodied-consciousnesses. For example, human consciousnesses.

3) Physical matter.

Physical matter of the current cosmos is just a daydream-stuff-composed reality which has been given birth through the instrumentality of daydreaming on the part of the ubiquitous & the infinite field of consciousness called cosmic space which, in turn or seriatim, is the expanded, distended, dilated or inflated form, version or state of bodiless and dimensionless consciousness of the creator, god or whatever of the current cosmos.

On account of being just a daydream-stuff-composed reality which has been given birth through the instrumentality of daydreaming on the part of the ubiquitous and the infinite field of consciousness called cosmic space aka expanded, distended, dilated or inflated form, version or state of bodiless and dimensionless consciousness of the creator, god or whatever of the current cosmos, the physical matter of the current cosmos is not an ageless, deathless, timeless, immortal or eternal thing in the manner of the ubiquitous and the infinite field of consciousness called cosmic space, aka expanded, distended, dilated or inflated form, version or state of bodiless and dimensionless consciousness of the creator, god or whatever of the current cosmos, and in the manner of the large medley of the individual consciousnesses of the current cosmos, for example human consciousnesses of the current cosmos.

The finesse, panache, virtuosity or wizardry to form or create the daydream-stuff-composed *physical matter* inside its field of consciousness called cosmic space by the creator, god or whatever of the current cosmos through the instrumentality of daydreaming on its part, is merely a tool at the disposal of the creator, god or whatever of the current cosmos in order to give rise to variety, diversity or heterogeneity inside its nondescript or featureless field of consciousness called cosmic space, nothing more nothing less.

Creator's or god's produced or generated daydream-stuff-composed *physical matter*; which is extant or present inside creator's or god's field of consciousness aka cosmic space and which provides variety, diversity or heterogeneity inside creator's or god's nondescript or featureless field of consciousness aka cosmic space; is labelled as material world or mineral world or physical world by human consciousnesses of the current cosmos.

Before it metamorphosed, transmuted or transformed itself into the ubiquitous and the infinite field of consciousness called cosmic space via the process of expansion, distension, dilation or inflation of its bodiless and dimensionless consciousness, the creator, god of whatever of the current cosmos, existed or occurred as an incredible or awe-inspiring and the only one of its kind or unique, timeless, bodiless and dimensionless consciousness of infinite intelligence and emotion.

This form, version or state of consciousness of creator, god or whatever of the current cosmos namely the form, version or state which is described as the incredible or awe-inspiring and the only one of its kind or unique, timeless, bodiless and dimensionless consciousness of the infinite intelligence and emotion, is designated as creator's or god's supra-transcendental form, version or state by human awarenesses of the current cosmos.

In its supra-transcendental form, version or state, creator's or god's consciousness is called supra-transcendental consciousness.

As said before, the ubiquitous and the infinite field of consciousness called cosmic space is nothing but the expanded, distended, dilated or inflated form, version or state of bodiless and dimensionless consciousness of creator, god or whatever of the current cosmos.

In other words, the ubiquitous and the infinite field of consciousness called cosmic space is nothing but the expanded, distended, dilated or inflated form, version or state of supra-transcendental consciousness, nothing more nothing less.

The supra-transcendental consciousness, conversely, vice versa, the other way around, or in reverse, is the incredible or awe-inspiring and the only one of its kind or unique, bodiless and dimensionless form, version or state of the ubiquitous and the infinite, 3-D or the three-dimensional field of consciousness of creator, god or whatever of the current cosmos.

And this ubiquitous and the infinite, 3-D or three-dimensional field of consciousness of creator, god or whatever of the current cosmos is called cosmic space by human consciousnesses of the current cosmos.

Thus, the being or the entity, designated as cosmic space by human awarenesses of the current cosmos is nothing but the 3-D or the three-dimensional field of consciousness of creator or god or whatever of the current cosmos.

The creator, god or whatever of the current cosmos, in its supra-transcendental form, version or state, exists or occurs as an incredible or awe-inspiring and the only one of its kind or unique, bodiless and dimensionless consciousness of infinite intelligence and emotion.

To restate.

The entity called the 3-D or three-dimensional cosmic space by human awarenesses of the current cosmos, has been created or formed by the incredible or awe-inspiring and the only one of its kind or unique, bodiless and dimensionless, supra-transcendental consciousness of the infinite intelligence and emotion aka creator, god or whatever of the current cosmos, through the process of expansion, distension, dilation or inflation of its bodiless and dimensionless consciousness in order to generate, create, produce or give rise to enough room, space or territory inside its bodiless and dimensionless consciousness of infinite intelligence and emotion so as to spatially or territorially accommodate its daydream-stuff-composed, 3-D or three-dimensional, *material cosmos* or *mineral cosmos* or *physical cosmos* of the present-day.

~*~*~*~*~

ADWAIT–VEDANTA & CONSCIOUSNESS-6

The incredible being designated as supra-transcendental consciousness has a large number of other names such as Brahm, Brahman, Parmatma, Parmeshwara, Ishwara, Om, Aum, Providence, Divine, Divinity, God, Creator, Maker etc. etc.

The incredible being called supra-transcendental consciousness was extant or present as such or as supra-transcendental consciousness prior to its transformation, transmutation, or metamorphosis into the current or contemporary cosmos.

And the current cosmos essentially consists of the following three ingredients.

1. Cosmic space.

2. Physical matter.

And last but not least,

3. A large assortment of individual or separate or discrete, embodied-consciousnesses, for example human consciousnesses.

The supra-transcendental existence or supra-transcendental being known as supra-transcendental consciousness has become metamorphosed into the current cosmos firstly, through the process of expansion, distension, dilation or inflation of its bodiless and dimensionless, consciousness so as to form the current 3-D or the three-dimensional, cosmic space inside its being namely its bodiless and dimensionless, consciousness and secondly, through the activity of daydreaming on its part so as to form the current 3-D or the three-dimensional, daydream-stuff-composed, physical matter inside its being.

The supra-transcendental form, version or state of the creator, god or whatever of the current cosmos is described as the incredible or awe-inspiring and the only one of its kind or unique, timeless, bodiless and dimensionless consciousness, awareness, sentience or mind of infinite intelligence and emotion.

The incredible existence or being, extant in the supra-transcendental realm or state or, extant in the supra-transcendental form, version or state has been addressed by a large number of names by human consciousnesses of the current cosmos, as per their whim, fancy, desire or liking. Some of these names have already been listed or written down earlier. However, it is sometimes also called supreme or absolute or, supreme or absolute consciousness, awareness, sentience or mind.

The supreme or absolute consciousness, awareness, sentience or mind of supra-transcendental realm or state is a unique being. It is the author, creator or progenitor of the current cosmos and also of all the future cosmoses of whatever form or shape and of whatever size or dimension plus of whatever variety, diversity or heterogeneity it fancies, wants, desire or dreams.

The current cosmos is an existing reality, albeit a daydream-stuff-composed reality, which is extant or present inside the expanded, distended, dilated or inflated form, version or state of the supra-transcendental consciousness. This expanded, distended, dilated or inflated form, version or state of the supra-transcendental consciousness is called cosmic space by human consciousnesses of the current cosmos.

The current cosmos, - which is a daydream-stuff-composed reality formed or created by the expanded, distended, dilated or inflated form, version or state of supra-transcendental consciousness called cosmic space - is being perceived and experienced at the present moment by the supra-transcendental consciousness i.e. cosmic space on one hand and human consciousnesses of the current cosmos on the other.

However, all the future cosmoses of the supra-transcendental consciousness's fancy, want, desire or dream exist inside this supra-transcendental consciousness in a latent or potential form or, as a latency, dormancy or potentiality, waiting to be made patent, manifest or noticeable to this supra-transcendental consciousness at the moment of its choosing.

In its original, native or supra-transcendental form, version or state, the supra-transcendental consciousness aka Supreme, Absolute, Brahm, Brahman, Parmatma, Parmeshwara, Ishwara, Om, Aum, Providence, Divinity, Creator, Maker, God or whatever of the current cosmos and of all the future cosmoses exists in an incredible or awe-inspiring form called the only one of its kind or unique, timeless, bodiless and dimensionless consciousness, awarenesses, sentience or mind of infinite intelligence and emotion.

What has been said above can be put in another

way.

What is called supra-transcendental consciousness or, supreme or absolute consciousness, awareness, sentience or mind, is in fact the original or native form, version or state of creator, god or whatever of the current cosmos and of all the future cosmoses.

In its original or native form, version or state, the supra-transcendental consciousness or, Supreme or Absolute consciousness, awareness, sentience or mind, exists, inheres or abides in an incredible or awe-inspiring form, version or state which is described as its incredible or awe-inspiring, the only one of its kind or unique, bodiless and dimensionless form, version or state.

To sum up.

The supra-transcendental consciousness, or, supreme or absolute consciousness, awareness, sentience or mind aka creator, god or whatever of the current cosmos and, creator, god or whatever of all the future cosmoses, exists, inheres or abides in an incredible or awe-inspiring and, the only one of its kind or unique, form, version or state which is described as timeless, bodiless and dimensionless form, version or state.

This incredible or awe-inspiring and, the only one of its kind or unique, timeless, bodiless and dimensionless form, version or state of the supra-transcendental consciousness or, supreme or absolute consciousness, awareness, sentience or mind is called its original, native or transcendental form, version or state.

Compared to its original, native or supra-transcendental or timeless, bodiless and dimensionless form, version or state; the creator, god or whatever of the current cosmos and the creator, god or whatever of all the future cosmoses also exists, obtains or occurs - from time to time - in another form, version or state which is called its physical, material or 3-D or three-dimensional form, version or state.

The physical, material or 3-D or three-dimensional form, version or state of creator, god or whatever is the current or the contemporary form, version or state of the latter.

The physical, material or 3-D or three-dimensional form, version or state of creator, god or whatever i.e. the current or the contemporary form, version or state of creator, god or whatever, is an elastic, plastic, puttylike or multifaceted form, version or state of creator, god or whatever. This is so because this elastic, plastic, puttylike or multifaceted form, version or state of creator, god or whatever i.e. the current or the contemporary form, version or state of creator, god or whatever, has three facets, faces, aspects, components or ingredients namely the cosmic space, the physical matter and last but not least, the large medley of individual, separate or discrete, embodied consciousnesses, for example human consciousnesses.

Thus, creator, god or whatever of the current or the contemporary cosmos and of all the future cosmoses exists, obtains or occurs in two forms, versions or states.

These two forms, versions or states of creator, god or whatever exist, obtain or occur in a periodic or rhythmic fashion.

They are as follows :-

Dimensionless form (Original or Native form).

Dimensional form (Metamorphosed or Transmuted form).

The dimensional form, version or state or better still, the 3-D or three-dimensional form, version or state of creator, god or whatever of the current or the contemporary cosmos and of all the future cosmoses is a metamorphosed or transmuted form, version or state of the original, native or supra-transcendental form, version or state of creator, god or whatever of the current or the contemporary cosmos and of all the future cosmoses.

The dimensional form, version or state or better still, the 3-D or the three-dimensional form, version or state of creator, god or whatever of the current or the contemporary cosmos and of all the future cosmoses is called its physical, material or mineral form, version or state.

The physical, material or mineral form, version or state of creator, god or whatever is latter's current or the contemporary form, version or state.

Inside physical, material or mineral or, better still, inside dimensional or 3-D or three-dimensional form, version or state of creator, god or whatever, as is the case at present, mankind and other embodied, dimensional or 3-D or three-dimensional but consciousness-endowed denizens or inhabitants on one hand and the embodied, dimensional or 3-D or three-dimensional but consciousness-lacking physical things or items such as planets, stars and galaxies on the other live, exist, prosper, make good, get ahead or make their mark.

By the way, all the embodied, dimensional or 3-D or three-dimensional but consciousness-endowed denizens or inhabitants such as human beings on one hand and the embodied, dimensional or 3-D or three-dimensional but consciousness-lacking physical things or items such as planets, stars and galaxies on the other, live, exist, prosper, make good, get ahead or make their mark inside the expanded, distended, dilated or inflated consciousness, awareness, sentience or mind of creator, god or whatever and nowhere else.

And this expanded, distended, dilated or inflated consciousness, awareness, sentience or mind of creator, god or whatever is none other than the entity called cosmic space by human awarenesses of the current or the contemporary cosmos.

As said before, the original, native or dimensionless form, version or state of creator, god or whatever of the current cosmos and of all the future cosmoses is called the supra-transcendental form, version or state of creator, god or whatever of the current or the contemporary cosmos and of all the future cosmoses.

The original, native, dimensionless or supra-transcendental form, version or state and the transmuted, metamorphosed, dimensional or 3-D or three-dimensional or physical form, version or state of creator, god or whatever of the current or the contemporary cosmos and of all the future cosmoses are two forms, versions or states or, are two aspects, manifestations or expressions of the same truth namely the creator, god or whatever of the current or the contemporary cosmos and all the future cosmoses.

The original, native, dimensionless or supra-transcendental form, version or state of creator, god or whatever of the current or the contemporary cosmos and of all the future cosmoses is described as the incredible or awe-inspiring and the only one of its kind or unique, timeless, bodiless and dimensionless consciousness, awareness, sentience or mind of infinite intelligence and emotion.

During its transmutation or metamorphosis into its physical, dimensional or 3-D or three-dimensional dimensional form, version or state namely the current or the contemporary cosmos, the original, native, dimensionless or supra-transcendental form, version or state of creator, god or whatever of the current or the contemporary cosmos and all the future cosmoses first expands, distends, dilates or inflates itself in order to form or create a room, space or territory inside its original, native, dimensionless or supra-transcendental form, version or state. This room, space or territory inside the original, native, dimensionless or supra-transcendental form, version or state of creator, god or whatever of the current or the contemporary cosmos and all the future cosmoses is called the cosmic space by human consciousnesses of the current or the contemporary cosmos.

Thus, the current or the contemporary 3-D or three-dimensional dimensional, cosmic space is nothing but the room, space or territory inside the expanded, distended, dilated or inflated consciousness, awareness, sentience or mind of creator, god or whatever of the current or the contemporary cosmos and all the future cosmoses.

The current or the contemporary, dimensional or 3-D or three-dimensional dimensional, cosmic space has been formed or created by the original, native, dimensionless or supra-transcendental form, version or state of creator, god or whatever of the current or the contemporary cosmos and of all future cosmoses, inside its consciousness, awareness, sentience or mind through the process of expansion, distension, dilation or inflation of its dimensionless consciousness, awareness, sentience or mind in order to provide enough room, space or territory for the spatial or territorial placement and existence of its current or the contemporary daydream-stuff-composed, 3-D or three-dimensional, physical, material or mineral cosmos, nothing more nothing less.

ADWAIT–VEDANTA & CONSCIOUSNESS- 7

The **dimensionless** version of **consciousness** of creator, god or whatever of the current cosmos is the **original** version of **consciousness** of creator, god or whatever of the current cosmos.

During the creation of current or contemporary cosmos, the **dimensionless** aka **original** version of **consciousness** of creator, god or whatever, first expanded, distended, dilated or inflated **itself,** to form or create inside **itself** the incredible or awe-inspiring, ubiquitous and the infinite **field of consciousness** called the current or contemporary **cosmic space,** in order to make available for use, enough room, space or territory for the spatial or territorial placement and existence of its subsequently produced, **daydream-stuff-composed,** *physical, dimensional,* or *3-D* or *three-dimensional creation* which consists of such **daydream-stuff-composed** *physical items* as planets, stars and galaxies etc. on one hand and **daydream-stuff-composed** *physical bodies* of such **embodied consciousnesses** of the current cosmos such as human beings etc. on the other.

To repeat.

The current or contemporary **cosmic space** is an incredible or awe-inspiring, ubiquitous and the infinite **field of consciousness** of creator, god or whatever of the current or contemporary cosmos, nothing more nothing less.

Or, the current or contemporary **cosmic space** is nothing but creator, god or whatever of the current or contemporary cosmos who has adopted or assumed a **3-D,** or a **three-dimensional** persona or personality in order to become the current or contemporary **3-D,** or **three-dimensional** cosmos.

All **physical matter,** whatever be its size, shape, idiosyncrasy, and location in the current or contemporary cosmos and which is being constantly perceived and experienced in the current or contemporary cosmos by all **human consciousnesses** - is merely a **condensed, compacted** or **compressed** *section* or *segment* of the incredible or awe-inspiring, ubiquitous and the infinite **field of consciousness** of creator, god or whatever of the current or contemporary cosmos.

And this ubiquitous and the infinite **field of consciousness** of creator, god or whatever of the current or contemporary cosmos is none other than the entity called the current or contemporary **cosmic space.**

The current or contemporary **cosmic space;** which is the incredible or awe-inspiring, ubiquitous and the infinite **field of consciousness** of creator, god or whatever of the current or contemporary cosmos; has **separated** a pure or pristine *section* or *segment* of itself from itself and out of this pure or pristine *section* or *segment* of itself, it has created, generated, or given rise to a large number of individual, separate or discrete, **dimensionless** but **embodied,** *secondarily consciousnesses* such as human consciousnesses etc.

Each and every such individual, separate or discrete, **dimensionless** but **embodied,** *secondary consciousness* such as human consciousness, which is extant or present inside the current or contemporary, ubiquitous and the infinite **field of consciousness** called **cosmic space,** is made to dwell, exist, live or reside inside an individual or distinct, **daydream-stuff-composed, dimensional,** or **3-D** or **three-dimensional** body called the **physical** or the **material** body.

By the way, the current or contemporary, ubiquitous and the infinite **field of consciousness** called **cosmic space** is the **dimensional** or, better still, is the **3-D** or the **three-dimensional** form, version or state of **consciousness** of creator, god or whatever of the current or contemporary cosmos which, in its **original** form, version or state, existed or occurred as an incredible or awe-inspiring, **bodiless** and **dimensionless consciousness** or, better still, as an incredible or awe-inspiring, **the only one of its kind** or unique **timeless, bodiless** and **dimensionless consciousness** of infinite intelligence and emotion.

All the **dimensionless** but **embodied** *secondary consciousnesses* such as human consciousnesses - which are extant or present inside the current or contemporary **cosmic space** i.e. inside the current or contemporary **dimensional,** or **3-D** or **three-dimensional** form, version or state of **consciousness** of creator, god or whatever of the current or contemporary cosmos - dwell, exist, live or reside inside an individual or distinct, **daydream-stuff composed, dimensional,** or **3-D** or **three-dimensional** body called the **physical** body and are therefore, classified into such groups as human, animal and plant, depending upon the distinctive, **idiosyncrasies** or **characteristics** of their individual or personal, **dimensional,** or **3-D** or **three-dimensional** body called the **physical** body.

Each and every individual, separate or discrete, **dimensionless** but **embodied,** *secondary consciousness* - which is extant or present inside the current or contemporary **cosmic space** i.e. which is extant or present inside the current or contemporary **dimensional,** or **3-D** or **three-dimensional** form, version or state of creator, god or whatever of the current or contemporary cosmos and each of which dwells, exists, lives or resides inside an individual or distinct, **daydream-stuff-composed, dimensional,** or **3-D** or **three-dimensional** body called the **physical** body - is ageless, deathless, timeless, eternal, immortal, inextinguishable or indestructible in the manner of its source, font or wellspring namely the creator, god or whatever of the current or contemporary cosmos.

To repeat.

Each and every individual, separate or discrete, **dimensionless** but **embodied,** *secondary consciousness* such as human consciousness - which is extant or present inside the current or contemporary **cosmic space** i.e. which is extant or present inside the current or contemporary

dimensional, or **3-D** or **three-dimensional** form, version or state of creator, god or whatever and, each of which dwells, exists, lives or resides inside an individual or distinct, **daydream-stuff-composed, dimensional,** or **3-D** or **three-dimensional** body called the **physical** body - is an absolutely authentic or genuine part, portion, segment or section of **the only one of its kind** or unique, **supra-transcendental consciousness** aka creator, god or whatever of the current or contemporary cosmos with regards to its intrinsic or inherent qualities including the quality of being immortal or eternal, nothing more nothing less, even though the fact remains that it dwells, exists, lives or resides inside an extremely transient, temporary or short-lived, **dimensional,** or **3-D** or **three-dimensional** physical or material body.

The **original** version or, **bodiless and dimensionless** version or, **supra-transcendental** version of creator, god or whatever of the current or contemporary cosmos has another name and this name is **primary consciousness.**

The use of the name **primary consciousness** for the **original** version or, for the **bodiless and dimensionless** version or, for the **supra-transcendental** version of creator, god or whatever of the current or contemporary cosmos, is very convenient or handy when one engages in contrasting this **primary**

consciousness with its **progenies** or **off-springs** namely the **secondary consciousnesses** such as human consciousnesses which exist or occur inside the current or contemporary, **dimensional** or **3-D** or **three-dimensional** form, version or state of **primary consciousness** namely the current or contemporary **cosmic space.**

All the **embodied** but **dimensionless,** *secondarily consciousnesses* such as human consciousnesses - which are extant or present inside the current or contemporary **cosmic space** i.e. which are extant or present inside the current or contemporary, **dimensional,** or **3-D** or **three-dimensional** form or version of creator, god or whatever - dwell, exist, live or reside inside their respective, individual or distinct, **daydream-stuff-composed,** dimensional or 3-D or three-dimensional body called the **physical** body.

What has been said above can be put in another way.

All the **embodied** but **dimensionless** *secondary consciousnesses* such as human consciousnesses of **physical, dimensional** or **3-D or three-dimensional** realm or domain - which dwell, exist, live or reside inside their respective, individual or distinct, **daydream-stuff-composed,** physical, dimensional, or 3-D or three-dimensional **physical** bodies - are absolutely real chunk or slice of bodiless and dimensionless, **primary consciousness** or **supra-**

transcendental consciousness, nothing more nothing less.

All the **dimensionless** but **embodied** *secondarily consciousnesses* such as human consciousnesses of physical, dimensional, or 3-D or three-dimensional realm or domain - which dwell, exist, live or reside inside their respective, individual or distinct, **daydream-stuff-composed,** dimensional, or 3-D or three-dimensional bodies called the **physical** bodies of the physical or dimensional realm or domain - are grouped together, under a single heading or banner called the **embodied consciousnesses** in order to contrast them with the **original** and **the only one of its kind** or unique, **non-embodied** or **bodiless** and **dimensionless**, **primary consciousness** of infinite intelligence and emotion who is their source, creator, maker, progenitor or whatever and also the source, creator, maker or progenitor or whatever of the entire current or contemporary cosmos.

HINDUISM (SANATAN DHARMA)

Sanatan Dharma or Hinduism is not a faith in the manner Christianity and Islam are.

It instead is a MAHA-BHANDAR or Super-Store of quest-tools from where human beings can acquire a tool as per their genetic make-up and as per subsequent evolution of their genetic make-up at any given point of the journey of their human-life, in order to understand the meaning of their own existence on one hand and the meaning of existence of the current cosmos in which they find themselves on the other.

Hinduism or Sanatana Dharma allows all human

beings total freedom at any given point of evolution of their genetic make-up, right through their lives to obtain new quest-tools from this MAHA-BHANDAR or Super-Store i.e. Hinduism or Sanatana Dharma, as they develop further in their quest.

Simultaneously, Hinduism is an AWARENESSBAL or CONSCIOUSNESSBAL SCIENCE in contrast to the material science of the present-day.

Since Hinduism or Sanatana Dharma is a science albeit an AWARENESSBAL or CONSCIOUSNESSBAL science, it is constantly evolving and growing and has been doing so from the time of its beginning some five thousand years ago. Thus, it is not a fossilised MAHA-BHANDAR or Super-Store of quest-tools. It instead is continually evolving and growing plus adding new quest-tools constantly into its MAHA-BHANDAR or Super Store, acquired through the AWARENESSBAL or CONSCIOUSNESSBAL experimentations and findings of the very long line of its seekers, searchers, researchers, bhaktas, sanyasies, swamies, gurus, acharyas, Sakkara-charyas, munnies, rishies, yogis, paramhansas and avatars.

As per Hinduism, the current cosmos is a transmuted or metamorphosed form of a bodiless, dimensionless SUPREME AWARENESS or SUPREME CONSCIOUSNESS which can be called by any name by human beings as per their whim, fancy, desire or liking.

The above view of Hinduism or Sanatana Dharma stands in total contrast to the view of material science of the present day. The material science of the present-day theorizes that the current cosmos is a transmuted or metamorphosed form of an infinitely hot, infinitely dense and infinitesimally small lump of physical matter called SINGULARITY or COSMIC EGG.

Lastly, one will like to add that both Christianity and Islam are one tool faiths which are accepted by their faithful on the basis of their absolute faith in it.

~*~*~*~*~

SUPRA-TRANSCENDENTAL EXISTENCE

Supra-transcendental existence is an incredible or awe-inspiring **consciousness, awareness, sentience** or **mind** and not any kind of physical matter.

This **supra-transcendental existence** or **supra-transcendental consciousness** is the creator, maker or progenitor of the current or contemporary cosmos.

Supra-transcendental existence or **supra-transcendental consciousness** is the name given to the **original form** or **version** of **consciousness** of creator, maker or progenitor of the current or the contemporary cosmos.

The other names of this **supra-transcendental existence** or **supra-transcendental consciousness** are **Saangyic consciousness** or **Brahmanic consciousness.**

As a matter of fact, this incredible or awe-inspiring **supra-transcendental, Saangyic or Brahmanic consciousness** or **original form** or **version** of **consciousness** of creator, maker or progenitor of current or contemporary cosmos can be called by any name by **human consciousnesses** as per their whim, fancy, desire or liking because it does not matter by what name this **original form** or **version** of **consciousness** of creator, maker or progenitor of the current or contemporary cosmos is called. However, what really does matter though is that **human consciousnesses** of the current cosmos must clearly understand that the **creator, maker** or **progenitor** of the current or contemporary cosmos is an incredible or awe-inspiring **consciousness** & not an infinitely hot, infinitely dense and infinitesimally small lump of **physical matter** called **singularity** or **cosmic egg** as theorised by the material scientists of the present-day.

Another name of supra-transcendental consciousness **is god.**

The **supra-transcendental, Saangyic or Brahmanic consciousness** or **original form** or **version** of **consciousness** of creator, maker or progenitor of the current or contemporary cosmos is described as an incredible or awe-inspiring and **the only one of its kind or** unique, **timeless, bodiless** and **dimensionless consciousness** of infinite intelligence and emotion.

HUMAN SLEEP AND ITS VITAL ROLE IN DECODING THE NATURE OF THE CURRENT COSMOS AND ITS CAUSE -1

SAANGYIC, BRAHMANIC OR SUPRA - TRANSCENDENTAL ANUBHUTI OR EXPERIENCE

Human consciousnesses are present in the current cosmos and presently at least, they seem to be the most intelligent of all the **consciousnesses** present in the current cosmos.

These **human consciousnesses** are provided, albeit for a very fleeting moment each night, in their **state of sleep,** a direct and real **Anubhuti, Experience** or

Taste of **Saangyic, Brahmanic, Supra-transcendental, Original** or **Bodiless & Dimensionless state** of creator, maker or progenitor of the current cosmos.

This direct and real **Anubhuti, Experience** or **Taste** of **Saangyic, Brahmanic, Supra-transcendental, Original** or **Bodiless and Dimensionless state** of creator, maker or progenitor of the current cosmos, which **human consciousnesses** of the current cosmos are provided during their **state of sleep** each night, even though it is of a very fleeing moment, yet it can be brought to **memory** in a very dynamic, graphic, lucid, powerful or vivid way by **human consciousnesses** during their **state of wakefulness,** provided they commit adequate **attention** to a particular **Bindu, Ekatva, Point** or **Unity** or, better still, to a particular **Sandhi, Sangam, Junction** or **Union** which is encountered by them during their **state of sleep** each night.

In **Adwait-Vedanta,** this **Bindu, Ekatva, Point** or **Unity** or, this **Sandhi, Sangam, Junction** or **Union ,** which is encountered for a very fleeting moment during their **state of sleep** each night by all **human consciousnesses,** is like, akin, near, or close to the **Bindu, Ekatva, Point** or **Unity** or, **Sandhi, Sangam, Junction** or **Union** at which the creation of the current cosmos began some 13.7 billion light years ago and will revert to same **Bindu, Ekatva, Point or Unity** or, **Sandhi, Sangam, Junction** or **Union** when the current cosmos will come to an end in an unknown future. This **Bindu, Ekatva, Point** or **Unity** or, this **Sandhi, Sangam, Junction** or **Union** can be said as being like, akin, near or close to the **"Consciousnessbal manifestation, representation or avatar of the current cosmos in its dormant, latent, potential or unmanifested form or version".**

The **Anubhuti, Experience** or **Taste** of this **Bindu, Ekatva, Point** or **Unity** or, this **Sandhi, Sangam, Junction** or **Union** by **human consciousnesses** which the latter acquire, albeit for a very fleeting moment, during their **state of sleep** each night, is called the **Saangyic, Brahmanic, Supra-transcendental, Original** or **Bodiless and Bodiless** *Anubhuti, Experience or Taste.*

To repeat.

The **Anubhuti, Experience** or **Taste** obtained by **human consciousnesses** at this incredible **Bindu, Ekatva, Point or Unity** or, **Sandhi, Sangam, Junction** or **Union** is called **Saangyic, Brahmanic, Supra-transcendental, Original** or **Bodiless and Dimensionless Anubhuti, Experience** or **Taste.**

Therefore, **human consciousnesses** of the **wakeful state** must not deal with the phenomenon of **human sleep** in a very **dismissive, casual** or **offhand way** and dedicate their best attention exclusively, solely or singularly to their **wakeful state** which is currently their wont or habit.

Of course, what has been said above, requires deeper explanation but before this can be provided, what is needed first is to discuss, explain and elaborate what **adult human sleep** itself truly is. This has been done in the next chapter.

HUMAN SLEEP AND ITS VITAL ROLE IN DECODING THE NATURE OF THE CURRENT COSMOS AND ITS CAUSE - 2

ADULT   HUMAN SLEEP

During sleep an individual is in its **constructive phase of metabolism** meaning thereby an individual's **body cells produce protoplasm for growth and repair** of the body's various systems, for examples :-

Immune system.

Nervous system

Skeletal system

Muscular system etc.

The internal **circadian rhythm** encourages daily sleep at *night* in **diurnal animals** such as human beings. In **nocturnal animals** such as **rodents,** on the other hand, the **internal circadian rhythm** encourages daily sleep in the *day*.

All sleep, even during the day, is associated with the secretion of the hormone **prolactin.**

Each stage of sleep, during each and every **sleep cycle** provides benefits to the sleeper but **deep sleep** and **dream sleep stages** are particularly beneficial.

Sleep increase an individual's sensory threshold to outside stimuli. As a result, sleeper perceives relatively few of them during sleep.

Human sleep occurs in **cycles** of about one and a half hour each. Every one of these **sleep cycles** consists of two **stages** called **non-REM** and **REM stages.**

There are 4 to 5   such **sleep cycles** each night.

Each of these **sleep cycles** of about one & a half hour each, have an increasing period of **REM sleep** and a decreasing period of **non-REM sleep** as these **sleep cycles** repeat themselves through the night. This is called **ULTRADIAN sleep cycle.**

The **cycles** labeled as being **ULTRADIAN** in mode,

have a period of recurrence shorter than a day but longer than an hour.

In contrast, an **INFRADIAN cycle** has a period of recurrence longer than the period of an **ULTRADIAN cycle.**

Examples of **INFRADIAN cycles** are seasonal cycles, breeding cycles etc.

Sleep is characterized by four criteria: reduced motor activity, reduced response to external stimuli, predictable posture (lying down with eyes closed), and a relatively ready reversibility. These features differentiate sleep from coma and hibernation.

Compared to **wakefulness** and **REM sleep, non-REM sleep** is characterized by electrical brain waves of **slower frequency** and **larger amplitude.** These electrical brain waves are recorded on an **electro-encephalographic** trace or **E.E.G** trace.

From the time sleeper falls asleep to the time he reaches deepest **stage** of his sleep namely the **non-REM sleep**, the frequency of these electrical brain waves on **E.E.G tracing** decreases progressively, while their amplitude increases correspondingly.

FOUR KINDS OF ELECTRICAL BRAIN WAVES

There are four **frequency ranges** and their corresponding **amplitude ranges** of electrical brain waves as seen on **E.E.G.** trace.

From the **highest** to the **lowest** frequency brain waves, the latter are divided into following four groups.

Beta waves: have a frequency range from **13** to **60** Hertz. (1 Hertz equals 1 oscillation per second) and an amplitude of about **30** microvolts.

Beta waves are the ones which are recorded on an **E.E.G trace** when human beings are **awake, alert** and **actively processing information.**

Sometimes the waves whose frequency is above **30** Hertz or **30** oscillations per seconds are labeled as **gamma waves. Gamma waves** are thought to indicate that **human consciousness** is establishing communication amongst various parts of the brain with the aim to formulate a ***coherent concept.***

Alpha waves: have a frequency range from **8** to **12** Hertz or **8** to **12** oscillations per seconds and an amplitude range of **31** to **50** microvolts.

Alpha waves are typically found in people who are awake but have their eyes closed and are relaxing or meditating.

Theta waves: have a frequency range of **4** to **7** Hertz and an amplitude range of **51** to **100** microvolts.

Theta waves are seen in **non-REM sleep stages** 1 and 2.

Delta waves: have a frequency range of **0.5** to **3** Hertz and an amplitude range of **101** to **200** microvolts.

Delta waves are seen when an individual is in **non-REM stage** 3 **sleep (deep sleep)** or in **coma.**

Flat-line E.E.G. trace: Lastly, when there is no electrical brain waves, the **E.E.G.** shows a **flat-line,** which is a clinical sign of **brain-death.**

The above described four kinds of electrical brain waves plus two more quite **specialized wave forms** (i.e. sleep spindles and K-complexes) as recorded on **E.E.G. trace,** constitute the vital criteria to divide normal **adult human sleep** into **four stages.**

Out of these **four sleep stages, three stages** belong to **non-REM sleep** and **one stage** to **REM sleep.**

FOUR STAGES OF ADULT HUMAN SLEEP

As said above, when each day, human beings go to sleep at night, their **sleep** becomes sub-divided into two types namely:-

1. Non-rapid eye movement sleep (non-REM sleep).

2. Rapid eye movement sleep (REM sleep).

In adult humans, sleep occurs in **repeating cycles.** Each such **sleep cycle** is made up of the above mentioned **two highly distinct stages,** namely the **non-REM** and **REM stages** which are called **non-REM** and **REM sleeps.**

Non-REM and **REM sleeps** are so different that they are regarded as **two** entirely **distinct states of human consciousness.**

As a matter of fact, **wakeful state, non-REM sleep state** and **Rem sleep state** are **three major states** of **human consciousness.**

A. NON-REM SLEEP (Non-Rapid Eye Movement Sleep)

During **non-REM sleep (non-Rapid Eye Movement Sleep)** brain uses significantly less amount of energy than in the **wakeful state.**

Brain restores its supply of chemical **ATP** or Adenosine-Tri-Phosphatase in those brain areas where brain activity has decreased during this type of sleep.

ATP is the chemical which brain uses for both short-term **storage** as well as short-term **transport** of energy in the brain.

THREE STAGES OF NON-REM SLEEP

The **non-REM sleep** is sub-divided into three **stages**:-

non-REM stage 1 sleep

non-REM stage 2 sleep

non-REM stage 3 sleep

Adult-human-sleep each night starts with **non-REM stage 1 sleep**, passes through **non-REM stages 2 and 3**, then back into **non-REM stage 2**, followed by **REM sleep** and finally ending with **non-REM stage 1 sleep** from where it started.

Obviously, the act of descending into deeper and deeper sleep is a gradual process but the above-mentioned sequence provide a convenient means of describing the relative depth of various **stages of non-REM sleep.**

non-REM stage 1 sleep: (Transition from wakefulness to sleep).

Non-REM stage 1 sleep constitutes about 5% of total sleep each night. Each period of this stage lasts about 3 to 12 minutes. Thus, this **sleep stage** is the briefest of all **sleep stages.**

Non-REM stage 1 sleep begins when the sleeper lies down and closes his eyes. Underneath sleeper's closed eyelids, his eye balls begin to move slowly. After the start of slow movements of eye balls underneath the closed eyelids, the transition to **non-REM stage 1 sleep** from wakefulness, occurs within few seconds to few minutes.

Then, as the sleep advances, **beta brain waves** of wakefulness are replaced by **alpha waves** of someone who is relaxed with his eyes closed. Soon **alpha waves** are followed by **theta waves** which are characteristic of **stage 1 sleep.**

Though the reaction of sleeper to external stimuli decreases during this **stage** of **sleep** but he is still in a very **shallow sleep.** Therefore, it is very easy to rouse him up from this **stage of sleep.** In experiments where people were awakened from **stage 1 sleep** and were then asked to describe their state of consciousness, they usually said that they had just fallen asleep or had been in the process of doing so.

During **non-REM stage 1 sleep** sudden muscular twitches, hypnic jerks, hypnagogic hallucinations and loss of most of awareness of wakeful state cosmos takes place.

Hypnic Jerk

A hypnic jerk also has a number of other names such as hypnagogic jerk, night start, sleep start or sleep twitch. It is an involuntary twitch which occurs just as a person is beginning to fall asleep. It often causes the person to awaken suddenly for a moment or two. Physically, hypnic jerks resemble the 'jump' experienced by a person when startled.

Hynogogic Hallucinations

During hypnagogic hallucinations person may hear sounds that are not there. He may also see things that are not there. These auditory and visual hallucinations are often very vivid. They may be very bizarre and/or disturbing.

1) non-REM stage 2 Sleep: (Light sleep)

Non-REM stage 2 sleep constitutes about 50% of total sleep each night. It is a state of light sleep. This stage of sleep helps in sleep-based, memory-consolidation as well as information-processing

E.E.G. during non-REM stage 2 sleep:

During this stage, *relatively slow frequency* theta waves predominate on E.E.G. trace. However, these theta waves are interrupted by occasional series of higher frequency waves known as *Sleep Spindles.* These sleep spindles have a frequency range of 8 to 14 Hertz and an amplitude range of 51 to 150 microvolts. Theta waves, on the other hand, have a frequency range of 4 to 7 Hertz and amplitude range of 51 to 100 microvolts.

Sleep spindles are generated due to interactions between thalamic and cortical neurons. Therefore, these sleep spindles are also called Thalmo-cortical spindles.

During stage 2 sleep, the E.E.G trace may also show a *wave form* called *K - Complex*.

A K - Complex is a biphasic wave form, consisting of a short negative deflection followed by a slower positive wave, and then a negative deflection once again. The duration of K- Complex is at least half a second.

K - complexes occur spontaneously and are thought to be evoked responses to internal stimuli.

Together, K - Complexes serve to protect sleep. They do this by suppressing sleeper's response to outside stimuli such as light, sound and touching of skin. Therefore, people in stage 2 sleep are unlikely to react to light or to noise unless it is extremely bright or loud. It is still possible to awaken these people. However, if they are awakened from stage 2 sleep, they report that they were really sleeping. This non-REM stage 2 sleep lasts 10 to 20 minutes during the earliest of night's sleep cycles.

Other things that happen during non-REM state 2 sleep:

During this stage of sleep, heart and respiratory rates both, slow down and body temperature drops so as to ready the body for the next stage of sleep called the non-REM Stage 3 sleep or deep sleep. Eye movements totally stop during this stage.

The non-REM stage 2 sleep the first stage of true sleep.

Duration of non-REM stage 2 sleep:

Sleepers pass through this stage of sleep more times than any other stage of sleep. As a result, they spend more time in stage 2 sleep than in any other stage. Typically, therefore stage 2 constitutes about 50% of total sleep time in adults.

non-REM stage 3 sleep: (Deep sleep)

This stage constitutes 25% of sleep each night and in adults it lasts for about 35 to 40 minutes during the first sleep cycle of the night.

Non-REM Stage 3 sleep, provides the deepest sleep to sleepers and so during it they sleep most soundly.

E.E.G in this stage of sleep is characterized by very slow frequency   and very high amplitude electrical brain waves called delta waves.

Non-REM stage 3 sleep is initiated in pre-optic area of the brain.

Neuronal activity is at its lowest. Blood flow is directed away from brain and pushed into muscles, restoring latter's physical energy. The muscles still maintain their muscle-tone, and some movements of the arm, legs and trunk are possible. Heart rate, blood pressure and respiratory rate are all decreased under the influence of parasympathetic nervous system. There is no eye activity. This is the stage of sleep during which most of the body's repair work is accomplished. During this stage, body not only repairs itself but also regrows tissues, builds muscles, and strengthens immune system. This sleep is the most restful sleep. It relieves the subjective feeling of sleepiness most. It restores the physical body most.

During non-REM stage 3 sleep, human secrete bursts of growth hormones. All sleep, even during the day, is associated with the secretion of hormone prolactin.

This is also the stage of sleep during which sleepers may suffer from what are called parasomnias. These include night terror, teeth grinding, sleep-eating, restless leg syndrome, nocturnal enuresis, sleep-walking (somn-ambulism or noct-ambulism) and sleep-talking (somni-loquy). During sleep-walking or somn-ambulism or noct-ambulism, sleep-walkers *arise* from *slow wave* or *delta wave sleep*, in a state of *low consciousness* and perform activities that are usually performed during a state of *full consciousness*. These activities can be as benign as sitting up in bed, walking to a bathroom, and cleaning, or as hazardous as cooking, driving, violent gestures, and grabbing at hallucinated objects.

Non-REM Stage 3 sleep, provides the deepest sleep & thus during it humans sleep most soundly. As a result, if someone wakes the sleeper up, he does not adjust immediately and often feels groggy and disoriented for several minutes.

During this stage of sleep, brain waves as seen on E.E.G. trace, are of very slow or low frequency and of very tall or high amplitude as compared to the REM sleep stage where brain waves are of high or rapid frequency and of small or low amplitude. These very slow or low frequency and very tall or high amplitude brain waves of non-REM Stage 3 sleep are called delta waves.

Since in non-REM stage 3 sleep brain waves are of slow frequency and high amplitude, this stage of sleep is also called slow-wave sleep in addition to being called delta-wave sleep.

Most non-REM stage 3 sleep occurs in the first half of the night. It therefore, occupies less time in the second half of the night. As a matter of fact, it may disappear altogether from the last sleep cycle. In contrast, REM sleep occurs mostly in the second half of the night.

The following features apply to some extent to all non-REM sleep stages

As sleeper descends into deeper and deeper **stages of sleep,** his body temperature, heart rate and respiratory rate all plummet, and every activity of his body slows down as does the body's energy consumption. Electrical brain waves become slower and slower in frequency and taller and taller in amplitude. As **non-REM sleep** progresses, brain becomes less and less responsive to outside stimuli. It thus, becomes more and more difficult to wake the sleeper from his sleep.

Excitatory neurotransmitter called *Acetylcholine* becomes less available in the brain.

However, reflexes remain fairly active.

B. REM Sleep (Rapid Eye Movement Sleep aka Dream Sleep state)

(a) Introduction

REM sleep constitutes about 20% of total sleep each night.

During REM sleep, the EEG trace looks much more like that observed in people who are awake and alert: electrical brain waves have a high frequency but their amplitude is small i.e. very much like that seen in awake and alert state. Hence this kind of sleep is called paradoxical sleep.

Thus, brain waves during REM Sleep, as seen on E.E.G. trace increase to the levels observed when a person is awake and alert.

What has been said above about REM sleep is described in more detail below.

REM sleep is called paradoxical-sleep firstly because in this state, sleeper, as per his E.E.G. trace, seems almost as *conscious* as he is in his wakeful state. His E.E.G. trace exhibits *high frequency and low amplitude* brain waves similar to that seen in the wakeful state, although paradoxically sleeper's consciousness is not aware of his physical body of which it was aware and will be once again aware in its wakeful state. Sleeper's consciousness in this state is not aware of the wakeful-state cosmos either, the wakeful state cosmos of which it was aware & will be once again aware when it will ascend into the wakeful state. Furthermore, while sleeper's brain, & other body systems such as respiratory system, cardiovascular systems etc. become much more active during dream sleep state, paradoxically his voluntary muscles of arm and legs are paralyzed (muscular-atony) for the period of REM SLEEP. This paralysis of voluntary muscles of sleepers is a protective-mechanism introduced by creator, god or whatever to protect the sleeper from self-harm and harm to others *through acting-out his dream or acting-out some scenes of his dream which often are very vivid, that is to say, which often are of such nature that they produce not only strong and clear images in the mind but also powerful feelings or emotions.* When one is paralyzed, one can't leap out of bed and run. Thus, during REM sleep, sleeper can breathe, his heart is beating, but he really can't move.

It has already been said before but it is worth repeating once again that electrical brain waves during REM sleep, as seen on E.E.G. trace, increase to the level experienced when a person is awake & alert.

An adult sleeper enters into REM sleep approximately every 90 minutes and remains in REM sleep far longer during the second half of night than during the first.

REM sleep starts as sleeper enters from non-REM stage 3 sleep into non-REM stage 2 sleep and thence into REM sleep.

Following the end of non-REM stage 3 sleep, a series of bodily movements signal the ascent of the sleeper into the lighter non-REM stage 2 sleep. This particular non-REM stage 2 sleep, following the non-REM stage 3 sleep, lasts only for about 5 to 10 minutes (as compared to 10 to 20 minutes following the non-REM stage 1), before sleeper enters into REM sleep stage.

The first REM sleep usually occurs about 70 minutes after one falls asleep & lasts about 10 minutes. Each of the subsequent REM sleep gets longer and longer, and final one may last up to an hour.

Non-REM - REM sleep cycles that occur during the first half of each night have a relatively long non-REM sleep and a relatively short REM sleep. In contrast, during the second half of each night, REM sleep stages lengthen and non-REM sleep stages shorten in duration.

There are 4 to 5 periods of REM sleep each night just as there are 4 to 5 sleep cycles each night.

If an individual has not had enough non-REM sleep, his body will try to make good this sleep, first at the expanse of REM sleep.

An adult sleeper spends about 20% of sleep in REM sleep each night.

Lack of REM sleep impairs one's ability to learn complex tasks when one is awake.

(b) Intense dreams during REM sleep

During REM sleep, an individual, dreams the most. These dreams tend to be crystal clear, very detailed, true to life and memorable. On account of this, REM sleep stage is also called a stage of active sleep.

If woken up during REM sleep, sleepers report that they had been dreaming graphically or vividly. Sometimes, these dreams are very bizarre. Sleepers are able to remember their dreams.

Dreams during "REM sleep" take as long as they actually seem.

Dreaming occurs during **REM sleep** because of increased brain activity. This increased brain activity can be seen objectively on **E.E.G. trace.**

(c) Electrical brain-waves as seen on E.E.G. trace during REM sleep

Electrical brain waves, as observed on E.E.G. trace, are increased to the level observed when a person is fully awake and alert. This E.E.G. or electro-encephalogram trace during REM sleep, is *paradoxically* much more like that observed in people who are awake. This *paradoxical* sleep is identified by *high frequency* and *low amplitude* electrical brain waves which are typical of an awake and alert brain.

REM sleep is called *paradoxical sleep* also because of one more reason which is that it is much harder to rouse sleepers from this stage of sleep than from any other stage of sleep. This is the situation despite the fact that sleeper's brain generates high frequency and low amplitude *electrical brain waves* similar to that seen in the cases of fully awake and alert state.

(d) Rapid eye movements in REM sleep

REM sleep is further characterized by numerous rapid eye-movements (REMs) that take place underneath sleeper's closed eyelids. Eyes move back and forth, that is to say, jerk rapidly under sleeper's closed eyelids. Hence, this state of sleep is referred to as REM sleep. These eye-movements are related in some ways to dreams which the sleeper perceives and experiences during this kind of sleep.

(e) Temporary paralysis of voluntary muscles of arms and legs during REM sleep

During REM sleep, muscles of arms and legs (voluntary muscles of the sleeper are temporarily paralyzed. This is a way to create a neurological barrier which prevents dreamer/sleeper from causing harm to himself and to others by acting out his dreams.

In REM sleep, the dreamer/sleeper can breathe, his heart beats but he can't move because of the temporary paralysis of all his four limbs.

(f) Other things that happen during REM sleep

In REM sleep, heart rate increases, blood pressure

rises and breathing rate is almost as much as in the wakeful state.

(g) Benefits of REM sleep

During REM sleep brain processes as well as consolidates the information sleeper has learned during the day time. Brain forms new neural connections. These new neural connections boost memory power. Brain also replenishes its supply of such neurotransmitters as serotonin and dopamine. These neurotransmitters boost mood during the day time. Thus, REM sleep is both memory-boosting as well as mood-boosting sleep for mankind.

REM sleep is turned on by those neurons which secrete neurotransmitter acetylcholine and is turned off by those neurons which secrete neurotransmitter serotonin.

Adult human sleep does not progress through all the four stages of sleep in sequence.

Let one explain what one means when one says that adult human sleep does not progress through all the four stages of sleep in sequence.

Sleep at night starts from non-REM stage 1 sleep. It then enters into non-REM stage 2 and thence, from non-REM stage 2, into non-REM stage 3.

Subsequently, however sleep reverts from non-REM stage 3 into non-REM stage 2.

Sleep there upon enters, from this repeat or second session of non-REM stage 2 of the same sleep cycle, into REM sleep stage.

What has been described above completes one full cycle of adult human sleep.

From REM sleep stage, sleep goes back into non-REM stage 1 from which place it began in the first place.

Then from this fresh non-REM stage1, a second cycle of adult human sleep commences.

The second cycle of sleep is completed in exactly the same manner as the first. In this way, all the sleep cycles of each and every night (usually about 4 to 5 in number) are brought to a close.

These 4 to 5 cycles of sleep per night are about 90 minutes each in adult human beings.

To reiterate.

Human sleep begins with non-REM stage 1 sleep. It then descends into non-REM stage 2. From non-REM stage 2, it descends further into non-REM stage 3 sleep.

After non-REM stage 3, human sleep returns to stage 2 sleep once again   before entering into the extraordinary realm of REM sleep. Finally, as just mentioned, sleep progresses from this repeat non-REM stage 2 into the mind-blowing REM sleep.

Thus, the sequence of human sleep is N1, N2, N3, N2, REM-sleep. (Here alphabet N stands for non-REM sleep). These 5 stages - during which stage N2 is entered into, partaken or savored twice - together constitutes one cycle of human sleep. On average total human sleep each night of 7 to 8 hours, is comprised of about 4 to 5 such sleep-cycles. And each of these sleep cycles consist of the sequence N1, N2, N3, N2, REM sleep. Then this cycle repeats itself once again till about 4 to 5 such sleep cycles are completed per night.

On average, each sleep-cycle take on average 90 minutes to complete but time is not absolutely fixed. Thus, a complete non-REM - REM cycle may require anywhere from 70 to 120 minutes to complete. First non-REM - REM cycle lasts about 70 - 100 minutes. Last non-REM - REM cycle lasts about 90 to 120 minutes.

End of REM sleep state

Once **REM sleep** is over, sleeper typically returns to **non-REM stage 1 sleep** once again in order to restart another **cycle of sleep** till about all 4 to 5 **sleep cycles** of the night in question are completed.

AWAKENING IN THE MORNING

After the completion of all the 4 to 5 **sleep cycles** of each and every **night,** sleepers ascend into the **wakeful state** in the morning. This they typically do via the **non-REM stage 1 sleep,** the **non-REM stage 1 sleep** into which they had ascended few minutes earlier from the last **REM sleep** of the night, that is to say, the **non-REM stage 1 sleep** into which sleepers had ascended few minutes earlier, after the completion of the **REM sleep** of the *last sleep cycle* of the night.

However, many a times, sleepers ascend into their **wakeful state** in the morning directly from the last **REM sleep** of the night, that is to say, directly from the **REM sleep** of the night's **last sleep cycle.**

HUMAN SLEEP AND ITS VITAL ROLE IN DECODING THE NATURE OF THE CURRENT COSMOS AND ITS CAUSE -3

DEEP SLEEP STATE

If human beings want to decode the true nature of the current cosmos; inside which they find themselves existing, living, and breathing totally unsought, unrequested or unsolicited; they must first understand clearly that their **deep sleep state** *encounter* plus their **dream sleep state** *experience* during their each and every **cycle** of **sleep** at night; affords them the greatest help with regards to this very vexed issue.

In respect of directly experiencing the **saangyic, brahmanic, supra-transcendental, original** or, **bodiless and dimensionless state** of **consciousness** of creator, maker, progenitor or whatever of the current cosmos, what is required of human consciousnesses is to provide adequate **attention** to a particular **point** or **vindu** of their **sleep state** each night or, if it is preferred, to a particular **junction, sangam** or **sandhi** of their **sleep state** each night. The direct experience which human consciousnesses acquire at this particular **point** or **vindu** of their **sleep state** each night or, at this particular **junction, sangam** or **sandhi** of their **sleep state** each night, is called the **saangyic anubhuti** or **experience** or, **brahmanic anubhuti** or **experience.** And this **anubhuti** or **experience** provides them the inkling or idea as to what **saangyic, brahmanic, supra-transcendental, original** or **bodiless** and **dimensionless consciousness** of creator, god or whatever of the current cosmos truly is.

Deep sleep state of human beings

A. What is human brain, human body, human consciousness, cosmic space and physical matter?

Human **brain** is the **antenna** via which a **pristine segment** of **consciousness** is **drawn in, sucked in** or **pulled in** into the human body from the surrounding, ubiquitous and the infinite **field of consciousness** called **cosmic space** which, in turn, is nothing but the expanded, distended, dilated or inflated form or version of the incredible or awe-inspiring and **the only one of its kind** or unique **bodiless and dimensionless consciousness** of creator, maker, progenitor or whatever of the current cosmos. Human **brain** also acts as a **storage depot** and **distribution center** for this **pristine segment of consciousness** from whence it is then distributed throughout the human body for this body's normal functioning.

The **pristine segment of consciousness,** drawn in, sucked in or pulled in by the **antenna** called **human brain** into the human body, from the surrounding, ubiquitous and the infinite **field of consciousness** called **cosmic space**, is designated as **human consciousness.**

1) What actually happens during deep sleep state of human beings?

During each episode of **deep sleep state**, the ability of the incredible **antenna** of the human body namely the **brain** to draw in, suck in or pull in a **pristine segment of consciousness** from the surrounding, ubiquitous and the infinite **field of consciousness** aka **cosmic space,** is temporarily curtailed or reduced by the **self-same,** ubiquitous and the infinite **field of consciousness** aka **cosmic space** in order to rest and repair the human body in question.

As a result, this **antenna** of the human body i.e. **the brain** is unable to draw in, suck in or pull in; during each and every **deep sleep** segment of **human sleep;** body's full, wakeful-state-allotted-quota of **consciousness** from the surrounding, ubiquitous and the infinite field of consciousness aka **cosmic space.**

Consequentially, the construct or the entity called **human being,** become totally unaware of its own existence which includes unawareness on its part not only of its physical body of which it was aware in its wakeful state but also the unawareness of existence of all its memories, thoughts, desires, and ambitions of which it was aware in its wakeful state.

It goes without saying that during its **deep sleep state,** this construct or entity called **human being** is also totally unaware of the existence of the **universe** of which it was aware in its **wakeful state.**

In short, the **deep sleep state** of a **human individual** is an incredible state of transient, temporary, evanescent, ephemeral, fleeting or flash **nothingness, nought, nada, zero, zilch, zippo, shoonya** or **cipher** from the perspective of that **human individual** i.e. **human individual** who is in its **deep sleep state.** It means, **albeit indirectly,** that in the **absolute absence** of **any kind of consciousness,** there will be **absolute absence** of **any kind of**

existence. In other words, the **Primary** or the **Primordial Existence** equals **Primary** or **Primordial Consciousness** and not the **Primary** or **Primordial physical matter** i.e. the **Singularity** or the **Cosmic Egg** of **Big Bang theory** fame, as currently theorized by the material scientists of today. **Physical matter** never was, never is and never will be the **Primary** or the **Primordial Existence.** It always was, always is and always will be a mere **secondary, subsidiary, subservient, non-essential** or **inessential existence,** created by the incredible **Primary** or **Primordial consciousness** aka the ubiquitous and the infinite **field of consciousness** i.e. **cosmic space** which, in turn, is nothing but the **expanded, distended, dilated** or **inflated** form or version of the incredible or awe-inspiring and **the only one of its kind** or unique, **timeless, bodiless** and **dimensionless consciousness** of creator, maker, progenitor or whatever of the current cosmos.

In the light of what has been said above vis-a-vis the **deep sleep state** of human beings, the latter namely, human beings must infer that their **incredible encounter** with their **deep sleep state** each and every night, should or rather must alert them to the fact of the **Primacy of Consciousness** in the affairs of the current cosmos and not the **Primacy of physical matter,** as is currently theorized by the material scientists of today via their famous **Big Bang theory** plus physical matter composed **Singularity** or **Cosmic Egg,** regarding the origin of the current cosmos.

The **physical matter** of the current cosmos is merely is a tool, vehicle or means of forming, creating or giving rise to **variety, diversity** or **heterogeneity** in the nondescript or featureless plus the ubiquitous and the infinite **field of consciousness** aka **cosmic space** which, in turn, is nothing but the expanded, distended, dilated or inflated form or version of the incredible or awe-inspiring and **the only one of its kind** or unique, **timeless, bodiless** and **dimensionless consciousness** of creator, maker, progenitor or god or whatever of the current cosmos.

The **physical matter** of the current cosmos is nothing but a **condensed, compressed** or **compacted** *section* or *segment* of the ubiquitous and the infinite **field of consciousness** aka **cosmic space** which, in turn, is nothing but the *expanded, distended, dilated or inflated* form or version of the incredible or awe-inspiring and **the only one of its kind** or unique **bodiless and dimensionless** *consciousness* of creator, maker, progenitor or god or whatever of the current cosmos.

The experience of **deep sleep state** is provided to human beings each night by creator, maker, progenitor or god or whatever of the current cosmos with a view to bring home or get across or convey or communicate to human beings the fact that

consciousness in the current cosmos is the **primary, supreme, ultimate** or **number one** *truth* and not the **physical matter.**

The **physical matter** in the current cosmos is merely a **secondary** construct or truth which has been created by **consciousness** or, more to the point, by the ubiquitous and the infinite **field of consciousness** aka **cosmic space** in order to create **variety, diversity** or **heterogeneity** in the nondescript or featureless, ubiquitous and the infinite **field of consciousness** aka **cosmic space, cosmic space** which, in turn, is nothing but the **expanded, distended, dilated or inflated** form or version of the incredible or awe-inspiring and **the only one of its kind** or unique **timeless, bodiless** and **dimensionless** *consciousness* of creator, maker, progenitor or god or whatever of the current cosmos.

From the perspective of an individual human being of the wakeful state of the current cosmos, he namely, this human being does not exist in his **deep sleep state.** This happens entirely due the fact that in **deep sleep state,** the amount of **consciousness** present in the human **brain** and therefore, present inside the rest of the human body is greatly diminished or curtailed by the ubiquitous and the infinite **field of consciousness** aka **cosmic space.** As a result, the construct or the entity called human being of the wakeful state; which is an extremely complex composite of all his extremely subtle and personal thoughts, memories, desires, ambitions, plans and pursuits, in addition to his gross body; does not exist in **deep sleep state.** In other words, despite the presence of gross human physical body in the **deep sleep state**, the **total construct** called human being of the wakeful state, ceases to exist in **deep sleep state.** This happens all because the supply of **consciousness** to the human **brain** and to the rest of the human body in **deep sleep state** is temporarily reduced or curtailed greatly during **deep sleep state.** Thus, the **encounter** with **deep sleep state** proves to the **construct** called human being of the wakeful state that it is the **consciousness** and **consciousness** alone or, more to the point, it is the ubiquitous and the infinite **field of consciousness** aka **cosmic space** and **cosmic space** alone in the current cosmos which is the gaffer or the guv or the chief or the **supreme** and not the **physical matter,** as currently theorized by the material scientists of today.

As it is, it is already very difficult for **physical matter** besotted, captivated or infatuated human beings of the wakeful state of the current cosmos, to convince themselves that the **consciousness** in the current cosmos is **primary** or **supreme** and not the **object of their love** or the **object of their undivided attention** called **physical matter.**

But without the existence each night of **deep sleep**

state and therefore, without the **encounter** each night with **deep sleep state,** it would have been absolutely impossible for human beings of the wakeful state of the current cosmos to convince themselves that the **consciousness** of the current cosmos or more to the point, the ubiquitous and the infinite **field of consciousness** aka **cosmic space** of the current cosmos, is the **primary** or **supreme truth** in the current cosmos and not the **physical matter.**

Although it is an absolutely impossible scenario, but let's hypothesize for a moment, for argument sake, the scenario where only **physical matter** exists and there exists no **consciousness** of any kind, not even the current **cosmic space.** In such a hypothetical scenario who will confirm or claim the existence of **insentient physical matter?** This is the vital question! The latter namely the **physical matter,** being an insentient construct or entity, is in no position to announce the truth of its own existence. Thus, in such a hypothetical scenario, the existence of **insentient physical matter** will remain unrecognized forever.

The above hypothetical scenario further illustrates the point that the **consciousness** or, more to the point, the ubiquitous and the infinite **field of consciousness** aka **cosmic space** is the **primary** or the **fundamental truth** in the current cosmos and not the **insentient physical matter.**

As said many times before, the ubiquitous and the infinite **field of consciousness** aka **cosmic space** is nothing but the **expanded, distended, dilated or inflated** form or version of the incredible or awe-inspiring and **the only one of its kind** or unique **timeless, bodiless and dimensionless *consciousness*** of creator, maker, progenitor or god or whatever of the current cosmos.

~*~*~*~*~

HUMAN SLEEP AND ITS VITAL ROLE IN DECODING THE NATURE OF THE CURRENT COSMOS AND ITS CAUSE- 4

DREAM SLEEP STATE

The **dream sleep state** is the other - (the first one being the **deep sleep state)** - extremely vital segment of each **cycle** of **human sleep** at night because it provides a very valuable insight **each night** about the true nature of the **wakeful state cosmos.** This insight can be had by **human consciousness** of the **wakeful state** provided the latter cares to analyze and explore the mystery of **dream sleep state** in depth.

Dream sleep state of human beings

1) What actually happens during dream sleep state of human beings?

The moment human beings enter into their **dream sleep state** from their **non-REM stage 2 sleep,** the **antenna** possessed by these individual's **physical body,** namely the **brain,** is permitted by the creator, god or whatever to **draw in, suck in** or **pull in almost** as much quantity of **consciousness** from the surrounding, ubiquitous and the infinite **field of consciousness** aka **cosmic space** as during their **wakeful state.** As a result, these individuals; even though they are in their **dream sleep state;** are almost as much **conscious** as they are during their **wakeful state.**

During **dream sleep state,** the electrical brain waves, as observed on the **E.E.G. trace,** increase to the level seen when a person is **awake** and **alert.**

In other words, the **E.E.G.** or the **electro-encephalographic trace** during **dream sleep state, is** *paradoxically* much more like that observed in people who are **awake. This** *paradoxical* **state of sleep** i.e. **dream sleep,** is identified by *low amplitude and high frequency* brain waves which are the characteristics of **awake** and **alert** human being.

The **dream sleep** of human beings is called

paradoxical sleep also because during it, the sleeper - even though he exhibits **low amplitude** and **high frequency** brain waves similar to that seen in the **wakeful state -** is much harder to rouse than during any other **sleep stage.**

B. Adwait-Vedantic perspective with regards to dream sleep state.

From **Adwait-Vedantic** viewpoint, several extremely vital phenomena make their mark or attain distinction during **dream sleep state** which are very important with regards to understanding the true nature of man's **wakeful state cosmos.**

Human consciousnesses give greatest importance to **wakeful state** not only because they exist in the **wakeful state** for much longer period than in the **sleep state,** but also because they believe the **wakeful state cosmos** to be **absolutely real** as compared, for example, to the **dream sleep state cosmos.** Therefore, it is very important that they must know whether the **wakeful state cosmos** is truly **absolutely real** as they believe it is, or it is as unreal as the **dream sleep state cosmos** which they perceive, experience, enjoy, hate and in which they take part with full vigor and interest during their **deep sleep state** each night by believing it to be **absolutely real** while they are perceiving, experiencing and taking part in it, even though when they wake up in the morning and become fully awake , they dismiss it as being nothing but a mere **dream** which appeared in their **consciousness** of the **dream sleep state** for some unknown reason and continues to do so each night during their **deep sleep state.**

The several extremely vital phenomena which make their mark or attain distinction during **dream sleep state** of human beings and which are very important with regards to understanding the true nature of man's **wakeful state cosmos** are described below.

Human consciousness of the **dream sleep state** is always totally unaware of the existence of the **wakeful state cosmos** of which it is **aware** in its **wakeful state.**

Human consciousness of the **dream sleep state** is also always totally unaware of the existence of its gross **physical body** of which it is **aware** in its **wakeful state.**

Thus, human consciousness in its **dream sleep state,** experiences an extraordinary **bodiless state** which is the hallmark or distinctive feature of the incredible or awe-inspiring and **the only one of its kind** or unique, **timeless** and **bodiless consciousness** of creator, maker or progenitor of the current cosmos.

What has been said above can be expressed in another way.

Human consciousness in its **dream sleep state** temporarily i.e. for the period of its **dream sleep state,** becomes an extraordinary **bodiless consciousness** in the manner of the incredible or awe-inspiring and **the only one of its kind** or unique, **timeless** and **bodiless consciousness** of creator, maker or progenitor of the current cosmos.

However, the extraordinary **bodiless** human consciousness of **dream sleep state** is not, and one will like to repeat here, is not a **dimensionless consciousness,** in the manner, creator, maker, or progenitor of the current cosmos was in its **original, supra-transcendental, saangyic,** or **brahmanic** form or version and will once again be in that form or version i.e. **dimensionless** form or version, when it will revert to its **original, supra-transcendental, saangyic,** or **brahmanic** form or version from its present or current **3-D** or **three-dimensional** form or version namely the ubiquitous and the infinite **field of consciousness** aka **cosmic space,** sometime in an unknown future as per its mood or fancy.

Instead, **each night,** the extraordinary **bodiless** human consciousness of **dream sleep state** is extant in its incredible **3-D** or **three-dimensional** form or version in the manner creator, maker or progenitor of the **current cosmos** is extant at the moment. By the way, in case one has forgotten, the incredible, **3-D** or

three-dimensional form or version of creator, maker or progenitor of the **current cosmos,** is presently extant as the ubiquitous and the infinite **field of consciousness** aka **cosmic space.**

It may be worthwhile here to reiterate what has been said above with regards to the present-day **cosmic space.**

The ubiquitous and the infinite **field of consciousness** aka **cosmic space** which human consciousnesses perceive and experience each day in their **wakeful state** is nothing but the **expanded, distended, dilated or inflated** form or version or the **3-D** or the **three-dimensional** form or version of the incredible or awe-inspiring and **the only one of its kind** or unique, **timeless, bodiless** and **dimensionless consciousness of** creator, maker or progenitor of the current cosmos.

Thus, in its **dream sleep state,** human consciousness temporarily becomes a **bodiless** but **3-D** or **three-dimensional consciousness** in the manner the ubiquitous and the infinite **field of consciousness** aka **cosmic space** of human's **wakeful state** is. The **cosmic space** of human's **wakeful state,** in turn, is nothing but the **expanded, distended, dilated or inflated** form or version or the **3-D** or **three-dimensional** form or version of incredible or awe-inspiring and **the only one of its kind** or unique **timeless, bodiless** and **dimensionless consciousness** of creator, maker or progenitor of the current cosmos. Furthermore, the **bodiless** but **3-D** or **three-dimensional human consciousness** of **dream sleep state** or the **bodiless** but **expanded, distended, dilated** or **inflated** form or version of **human consciousness** of **dream sleep state** perceives and experiences as detailed and varied a **dream world** as it perceives and experiences in its **wakeful state** but to this latter **dream word** i.e. the **dream world** it perceives, experiences and takes part in its **wakeful state** namely the **wakeful state cosmos**, it **ignorantly** or **nesciently** apportions, allots or handouts the label, tag or epithet of an absolutely **real world** or **real cosmos.**

However, the truth is that **both worlds** or **both cosmoses** i.e. the **dream sleep state world** or **cosmos** on one hand and the **wakeful state world** or **cosmos** on the other are mere **dreams** only and nothing else, albeit one amongst them i.e. the **dream sleep state cosmos** is a **dream** of the **dream sleep state** human consciousness whereas the other namely, the **wakeful state cosmos** is the **dream** of the ubiquitous and the infinite **field of consciousness** aka **cosmic space** which, in turn, is nothing but the **expanded, distended, dilated or inflated** form or version or the **3-D** or the **three-dimensional** form or version of creator, maker or progenitor of the current cosmos.

Inside the **human consciousness,** during its **dream sleep state,** a **consciousnessbal space** or a **cosmic space** appears or becomes visible, willy nilly or perforce under the influence of the **mighty will** of creator, god or whatever of the current cosmos. This happens, occurs or takes place during each and every episode of **dream-sleep** of human consciousness. It involves the **expansion, distention, dilation** or **inflation** of human consciousness during its **dream sleep state.** This process of **expansion, distention, dilation or inflation** of human consciousness during its each and every **dream sleep state** is the same process by which the **original, supra-transcendental, saangyic, brahmanic** or **dimensionless** form or version of **consciousness** of creator, maker or progenitor of the current cosmos has created inside itself, the present, ubiquitous and the infinite **field of consciousness** aka **cosmic space** which is perceived and experienced by all **human consciousnesses** when they are extant in their **wakeful state.**

Similarly the very realistic, solid-looking and **physical-looking objects** of the **dream sleep state world** or the **dream sleep state cosmos** such as mountains, rivers and stars etc. as well as the live and conscious physical bodies of human beings and animals etc. - which **human consciousness** perceives and experiences in its **dream sleep state** and which become visible and experienceable to it inside its own **consciousnesses** willy nilly or perforce under the influence of the **mighty will** of creator, god or whatever of the current cosmos - are nothing but a **condensed, compressed** or **compacted** segment of **dreamer's consciousness** or, better sill, are nothing but a **condensed, compressed,** or **compacted** segment of **dreamer's field of consciousness** aka **dreamer's consciousnessbal space** or **cosmic space,** all of which have appeared or have bobbed up **involuntarily** or **perforce** under the influence of the **mighty will** of creator, god or whatever of the current cosmos, inside **dreamer's consciousness only** and nowhere else during **dreamer's dream sleep state** and which are all, therefore, as said before, extant inside **dreamer's consciousness only** and nowhere else during **dreamer's dream sleep state.** All this has happened, as said before, under the influence of the **mighty force of will** of creator, god or whatever of the current cosmos.

It will be worthwhile to repeat what has been said above.

Similarly the very realistic, solid-looking and **physical-looking objects** of the **dream sleep state world** or the **dream sleep state cosmos** such as mountains, rivers and stars etc. as well as the live and conscious physical bodies of human beings and animals etc. - which **human consciousness** perceives and experiences in its **dream sleep state** and which become visible and experienceable to it inside its own **consciousness** willy nilly or perforce under the influence of the **mighty will** of creator, god or whatever of the current cosmos - are nothing but a **condensed, compressed** or **compacted** segment of **dreamer's consciousness** or, better sill, are nothing but a **condensed, compressed,** or **compacted** segment of **dreamer's field of consciousness** aka **dreamer's consciousnessbal space** or **cosmic space,** all of which have appeared or have bobbed up **involuntarily** or **perforce** under the influence of the **mighty will** of creator, god or whatever of the current cosmos, inside **dreamer's consciousness only** and nowhere else during **dreamer's dream sleep state** and which are all therefore, as said before, extant inside **dreamer's consciousness only** and nowhere else during **dreamer's dream sleep state.** All this has happened, as said before, under the influence of the **mighty force of will** of creator, god or whatever of the current cosmos.

The **dream world** or the **dream cosmos;** which is perceived and experienced by the **dream sleep state** human consciousness during its **dream-sleep** at night involuntarily or perforce under the influence of the **mighty force of will** of creator, god or whatever of the current cosmos; appear to the **dream sleep state** human consciousness as **real, solid** and **physical** as the **wakeful state cosmos** appears to the

wakeful state human consciousness. However, since the human consciousness in its **wakeful state** possesses the insight regarding the true nature of its **dream sleep state cosmos,** it has no hesitation in declaring that the **dream sleep state cosmos** was a mere **dream** and nothing but a mere **dream,** created involuntarily, willy nilly or perforce inside its **dream sleep state consciousness** by no less a being than the creator, god or whatever of the current cosmos.

Similarly, when one day the **wakeful state** human consciousness will develop true insight regarding the **real nature** of its **wakeful state cosmos,** it will then have no hesitation in declaring that even the **wakeful state cosmos** which it perceives , experiences and takes part in its **wakeful state** is also a mere **dream cosmos** or **dream world,** created by the selfsame creator, maker, god or whatever who creates each night the incredible **dream sleep state cosmos** inside the human consciousness of **dream sleep state.**

In other words, when one day, the **wakeful state** human consciousness will develop true insight regarding the **real nature** of the **wakeful state cosmos,** it will then have no hesitation in declaring that even the **wakeful state cosmos** is no more real than the **dream sleep state cosmos.** And the creator, maker or progenitor of the first or No. 1 **dream-stuff composed** i.e. the **wakeful state cosmos** is the same creator, maker or progenitor who creates each night the second or No. 2 **dream-stuff composed cosmos** inside each and every human consciousness when the latter is in its **dream sleep state.**

The only difference that obtains between the two, namely, the **dream sleep state cosmos** i.e. the second or No. 2 **cosmos** on one hand and the **wakeful state cosmos** i.e. the first or No. 1 **cosmos** on the other is that, whereas **the former** is created by creator, maker, god or whatever inside the **consciousness** of a human being when this human being descends into his **dream sleep state, the latter** is created by this creator, maker, god or whatever inside its own incredible or awe-inspiring, **the only one of its kind** or unique, **timeless** and **bodiless consciousness.** And, this **timeless** and **bodiless consciousness** of this incredible creator, maker, god or whatever, currently exists in its **expanded, distended, dilated or inflated** form or version or, **3-D** or **three-dimensional** form or version called the ubiquitous and the infinite **field of consciousness** aka **cosmic space.**

As said before, the **dream sleep state cosmos** which exists inside the **consciousness** of human beings when the latter are in their **dream sleep state,** is created by the selfsame creator, maker or progenitor who also is the creator, maker or progenitor of the current **wakeful state cosmos.**

Incidentally, the phrase "**wakeful state cosmos**" refers to the **cosmos** which **human consciousnesses** perceive, experience and take part in their **wakeful state.**

The **dream sleep state cosmos (**which exists inside the **consciousness** of human beings when the latter are in their **dream sleep state** and which is created by the selfsame creator, maker or progenitor who also is the creator, maker or progenitor of the current **wakeful state cosmos)** contains all the ingredients which also exist in the **wakeful state cosmos.** These ingredients are as follows.

Cosmic space.

Physical matter.

Consciousness. For example, human consciousness, animal consciousness etc.

And last but not least,

A perceiver and experiencer of this dream sleep state cosmos. This perceiver and experiencer of the dream sleep state cosmos is the human consciousness of dream sleep state who involuntarily, willy nilly or perforce perceives and experiences this dream sleep state cosmos inside its consciousness, the dream sleep state cosmos which consists of all the above listed ingredients and which has been created inside human consciousness of dream sleep state by the selfsame creator, maker or progenitor who is also the creator, maker or progenitor of the current wakeful state cosmos.

Here it will be worthwhile repeating once again the following.

The incredible or awe-inspiring conscious being who creates the dream sleep state cosmos inside human consciousnesses, when the latter descend into their dream sleep state, is the selfsame creator, maker or progenitor who has created the current wakeful state cosmos. And, incidentally, the current wakeful state cosmos is the cosmos which is currently being perceived and experienced by all human consciousnesses in their wakeful state.

Thus, the creator, maker or progenitor of both cosmoses namely the wakeful state cosmos on one hand and the dream sleep state cosmos on the other, is the selfsame or one and the same, incredible or awe-inspiring and the only one of its kind or unique timeless and bodiless consciousness of infinite intelligence and emotion called god, creator, maker, progenitor or whatever.

In other words, the dream sleep state cosmos is

created inside the consciousness of its perceiver & experiencer namely the human consciousness of the dream sleep state, by the very same creator, maker or progenitor who is also the creator, maker or progenitor of the current wakeful state cosmos.

The presence or existence of twin panoramas called the wakeful state cosmos on one hand and the dream sleep state cosmos on the other (both of which are perceived and experienced by human consciousnesses without fail every day and in which they take part without fail every day), is for the sake or purpose of providing a very powerful hint or clue to all human consciousnesses of the wakeful state that, contrary to popular belief, there is no genuine free will for anyone who exists either in the wakeful state cosmos or in the dream sleep state cosmos.

It ought to be absolutely evident to every human consciousness which is extant in the wakeful state cosmos, that since the dream sleep state cosmos is not real, there is no question of possession of a genuine free will on the part of any being of the dream sleep state cosmos. This includes the live and conscious human beings of the dream sleep state cosmos who, on most occasions, are present in the dream sleep state cosmos.

It also ought to be absolutely evident to every human consciousness which is extant in the wakeful state cosmos that the dream sleep state cosmos, even though present inside the human consciousness of the dream sleep state, is not created by the latter, that is to say, is not created by the human consciousnesses of dream sleep states. It instead, that is to, the dream sleep state cosmos instead, takes birth willy nilly, perforce or involuntarily inside the human consciousness of the dream sleep state. Therefore, there is no question of possession of a genuine free will on the part of the human consciousness of the dream sleep state either.

Exactly similar is the situation with respect to the wakeful state cosmos. Here too, no genuine free will exists for any being including all the live and conscious beings of the wakeful state cosmos such as human beings. The feeling which all the live and conscious human beings of the wakeful state cosmos harbor, nurse or entertain inside them that they definitely possess a genuine free will, is merely an inner mirage of theirs and nothing else, an inner mirage which has been embedded or planted deep inside them or, better still, embedded or planted deep inside their consciousness or mind by no less a being than the creator, maker or progenitor of the above mentioned both panoramas namely the wakeful state cosmos on one hand and the dream sleep cosmos on the other.

All beings, things, events, phenomena and activities that exist, take place or occur in both cosmoses, exist, take place or occur as per the mighty force of will of the incredible or awe-inspiring and the only one of its kind or unique, timeless & bodiless consciousness of creator, maker or progenitor of the current twin panoramas called the wakeful state cosmos on one hand and the dream sleep state cosmos on the other.

What has been said above can be put in another way.

The dream sleep state cosmos, even though it exists inside the consciousness of its perceiver and experiencer; is not actually created by this perceiver and experiencer. It instead, is created by a higher being. This higher being is none other than the entity called the creator, maker or progenitor of the current twin panoramas called the wakeful state cosmos on one hand and dream sleep state cosmos on the other.

Since the dream sleep state cosmos, even though it exists inside the consciousness of its perceiver and experiencer; is not actually created by its perceiver and experiencer and is instead, created by a higher being and since this higher being is none other than the entity who is the creator, maker or progenitor of the current twin panoramas called the wakeful state cosmos on one hand and the dream sleep state cosmos on the other, the creator, maker or progenitor of both these current panoramas is providing a powerful clue or hint to all human consciousnesses of the wakeful state, that, contrary to popular belief, there is no genuine free will for anyone even in the wakeful state cosmos just as there is no genuine free will for anyone in the dream sleep state cosmos.

In other words, just as the perceiver and experiencer of the dream sleep state cosmos as well as all the live and conscious beings such as human beings extant in the dream sleep state cosmos have no genuine free will, similarly all the live and conscious beings such as human beings extant in wakeful state cosmos have no genuine free will.

All beings, things, events, phenomena and activities that exist, occur or take place in both cosmoses namely, the dream sleep cosmos on one hand and the wakeful state cosmos on the other, exist, occur as take place as per the mighty force of will of the incredible or awe-inspiring and the only one of its kind or unique, timeless and bodiless consciousness of creator, maker or progenitor of the current twin panoramas called the wakeful state cosmos on one hand and the dream sleep state cosmos on the other.

HUMAN SLEEP AND ITS VITAL ROLE IN DECODING THE NATURE OF THE CURRENT COSMOS AND ITS CAUSE -5

THE NITTY-GRITTY OF SAANGYIC, BRAHMANIC OR SUPRA - TRANSCENDENTAL ANUBHUTI OR EXPERIENCE

It will be worthwhile here to go over again what was said earlier with regards to **Saangyic, Brahmanic** or **Supa-transcendental Anubhuti** or **Experience.**

Human consciousnesses are present in the current cosmos and presently at least, they seem to be the most intelligent of all the **consciousnesses,** present in the current cosmos.

These **human consciousnesses** are provided, albeit

for an extremely fleeting moment each night, in their **state of sleep,** a direct and genuine **Anubhuti, Experience** or **Taste** of **Saangyic, Brahmanic, Supra-transcendental, Original** or **Bodiless & Dimensionless state** of creator, maker or progenitor of the current cosmos.

This direct and genuine **Anubhuti, Experience** or **Taste** of **Saangyic, Brahmanic, Supra-transcendental, Original** or **Bodiless and Dimensionless state** of creator, maker or progenitor of the current cosmos, which **human consciousnesses** of the current cosmos are provided during their **state of sleep** each night, even though is of a very fleeing moment, it nevertheless can be brought to **memory** in a very dynamic, graphic, lucid, powerful or vivid way by **human consciousnesses** during their **state of full wakefulness,** provided they commit adequate **attention** to a particular **Bindu, Ekatva, Point** or **Unity** or, better still, to a particular **Sandhi, Sangam, Junction** or **Union** which is encountered by them during their **state of sleep** each night.

In **Adwait-Vedanta,** this **Bindu, Ekatva, Point** or **Unity** or, this **Sandhi, Sangam, Junction** or **Union ,** which is encountered for a very fleeting moment by all **human consciousnesses** during their **state of sleep** each night, is like, akin, near, or close to the **Bindu, Ekatva, Point** or **Unity** or, **Sandhi, Sangam, Junction** or **Union** at which the creation of the current cosmos

316

began some 13.7 billion light years ago and will revert to same **Bindu, Ekatva, Point or Unity** or, **Sandhi, Sangam, Junction** or **Union** when the current cosmos will come to an end sometime in the future. This **Bindu, Ekatva, Point** or **Unity** or, this **Sandhi, Sangam, Junction** or **Union** can be said as being like, akin, near or close to the **"manifestation** or **representation of the current cosmos in its dormant, latent, potential or unmanifested form or version".**

The **Anubhuti, Experience** or **Taste** of this **Bindu, Ekatva, Point** or **Unity** or, of this **Sandhi, Sangam, Junction** or **Union,** which the **human consciousnesses** of the current cosmos are provided during their **state of sleep** each night albeit for an extremely fleeting moment, is called the **Saangyic, Brahmanic, Supra-transcendental, Original** or **Bodiless and Dimensionless** *Anubhuti, Experience* or *Taste.*

To repeat.

The **Anubhuti, Experience** or **Taste** obtained by **human consciousnesses** at this incredible **Bindu, Ekatva, Point or Unity** or, at this incredible **Sandhi, Sangam, Junction** or **Union** is called **Saangyic, Brahmanic, Supra-transcendental, Original** or **Bodiless and Dimensionless Anubhuti, Experience** or **Taste.**

Therefore, **human consciousnesses** of the **wakeful**

state must not deal with the phenomenon of **human sleep** in a very **dismissive, casual** or **offhand way** and dedicate their best attention merely or exclusively to their **wakeful state** which is currently their wont or habit. Instead, they must pay some serious heed to this phenomenon of **human sleep** if they wish to unravel the true story of the current cosmos and its creator.

So, let's start.

When **human consciousnesses** traverse their various **sleep stages** as well as their various **sleep cycles** each night, there comes a time when they all are ready to **fully wake up.** This **fully waking up** process of **human consciousnesses** from their **night sleep** in the morning each day involves their **movement,** that to say, involves the **movement** of these **human consciousnesses,** either from their **non-REM stage 1 sleep** into the **fully awake state** or from their **REM sleep (dream sleep state)** into the **fully awake state.** Either way, the **Bindu, Ekatva, Point or Unity** or, **Sandhi, Sangam, Junction** or **Union** between the **human sleep state** on one hand and the **human fully awake state** on the other, is an extremely **vital confluence** or **sangam** or, if one likes, is an extremely **vital junction** or **sandhi.**

It is this **confluence, sangam, junction** or **sandhi** which forms or constitutes the **vindu** or **point** or, **ekatva** or **unity** from which flows the **clear understanding** for mankind, regarding that incredible or awe-inspiring form or version of creator, maker of progenitor of the current cosmos which existed when this creator, maker or progenitor was in its **original** form or version, that is to say, was in its **bodiless** and **dimensionless** form or version or, if it is preferred, was in its **supra-transcendental, saangyic,** or **brahmanic** form or version.

What has been said above requires further clarification.

At the above mentioned **vindu** or **point** or, if it is preferred, at the above-mentioned **confluence** or **sangam** of **human sleep state** on one hand and of **human fully awake state** on the other, a number of phenomena become patent to **human consciousnesses,** albeit for an extremely fleeting moment. These phenomena are described below.

Firstly, at the above mentioned **vindu, point, confluence** or **sangam** of **human sleep state** on one hand and **human fully awake state** on the other, all **human consciousnesses,** without fail, are **aware** or **conscious** of themselves as an **aware** or **conscious being,** in the manner they all were aware or conscious a little while earlier during their **REM sleep** or **dream sleep state** but with one extremely vital, key or all important difference between the two of them.

And the above mentioned one extremely vital, key or all important difference between the two of them is that, while during their **REM sleep** or **dream sleep state** all human consciousnesses or awarenesses are aware or conscious of the existence of a **dream sleep state cosmos** inside their consciousnesses or awarenesses, in stark contrast, no such **dream sleep cosmos** exists inside the **human consciousnesses or awarenesses** which are extant or present at the above mentioned **vindu, point, confluence** or **sangam.**

In all the other aspects though, all **human consciousnesses** or **awarenesses** which are extant or present at the above mentioned **vindu, point, confluence** or **sangam** are akin, alike, near or close to the **human consciousnesses** which are extant or present in the **dream sleep** or **REM sleep state,** meaning thereby, just like all the **human consciousnesses** which are extant or present in the **dream sleep** or **REM sleep state,** all the **human consciousnesses** which are extant or present at the above mentioned **vindu, point, sangam** or **confluence,** are neither **aware** or **conscious** of their **physical body** of which they all were **aware** or **conscious** in their **fully awake state** nor are they **aware** or **conscious** of the **physical cosmos** of which they all were **aware** or **conscious** and will be once again **aware** or **conscious** in their **fully awake** or **fully conscious state.**

It will not be inappropriate at this juncture to compare the three very important states of **human consciousnesses** against each other. These three states are called the **saangyic** or **brahmanic state,** the **dream sleep** or **REM sleep state** and, last but not least, the **fully awake** or **wide-awake state** of human consciousnesses.

Incidentally, the label " **saangyic** or **brahmanic** human consciousnesses" is affixed, applied or afforded to human consciousnesses that exist or are extant at the **vindu, point, confluence** or **sangam** of **human sleep state** on one hand and **human fully awake state** on the other.

The purpose of comparing the above mentioned three very important states of **human consciousnesses** is firstly to underline, highlight or stress the differences that obtain or exist between the three of them and then to underscore, highlight or stress the point of similarity that obtain or exist between the three of them.

To start with, it will be worthwhile to point out the one and only point of similarity that obtains or exists between all these three states of **human consciousnesses.**

This one and only point of similarity that obtains or exists between all these three states of human consciousnesses, consists of the fact that during all these three states of theirs, all human consciousnesses are **aware** or **conscious** of themselves as an **aware** or **conscious being.**

So, from the point of view of being **aware** or **conscious** of their own existence as an **aware** or **conscious being,** all these three states of **human consciousnesses** are quite alike, similar, close or near.

However, all these three states of human consciousnesses, without exception, are absolutely different from the one and only **deep sleep state** or the **non-REM stage 3, sleep-state** of mankind.

During the one and only **deep sleep state** or the **non-REM stage 3, sleep-state** of mankind, there exists or obtains not enough **human consciousness** inside the **physical brain** of the live **human body.**

As a result of the **physical brain** of the live **human body** not having enough amount of **human consciousness** during the **deep sleep state** or **non-REM stage 3, sleep-state,** these **human consciousnesses** are not **self- aware** in the usual way or, better still, loose their customary level of **self-awareness** and thus find themselves not in a position to proclaim their existence as **aware** or **conscious beings** in the usual way.

What has been said above can be put in another

way.

During the **deep sleep state** or **non-REM stage 3, sleep-state** of human beings, the amount of **human consciousness** extant or present inside the live **human brain** is so enormously curtailed that it becomes deprived of its customary level of **self-awareness.**

Now let's talk about the various differences that obtain or exist amongst the **saangyic** or **brahmanic state,** the **dream sleep** or **REM sleep state** and, last but not least, the **fully awake** or **wide-awake state** of human consciousnesses.

The **fully awake** or **wide-awake** human consciousnesses are aware or conscious of the existence of their **physical body** of the **fully awake** or **wide-awake** state. This is not the case either with **saangyic** or **brahmanic state**-human consciousnesses or with **dream sleep** or **REM sleep state**-human consciousnesses.

Similarly, **fully awake** or **wide-awake** human consciousnesses are aware or conscious of the existence of the **fully awake** or **wide-awake-state** cosmos of which neither the **saangyic** or **brahmanic state**-human consciousnesses nor the **non-REM stage 3 sleep-state** or the **dream sleep state**-human consciousnesses are aware of themselves.

Now let's deal once again with the **saangyic** or **brahmanic** human consciousnesses that exist or are extant every day without fail, albeit for a very short moment, at the above mentioned **vindu, point, confluence or sangam** of the human **sleep state** on one hand and the human **fully awake** or **wide-awake state** on the other.

It must be pointed out here that this incredible **vindu, point, confluence** or **sangam** of the human **sleep state** on one hand and the human **fully awake** or **wide-awake state** on the other must be **consciousnessbally** visualised in its "larger than life" form with the aid of the "magnifying lens" of the fully awake or wide-awake human consciousnesses if the latter are interested in understanding the **primordial** or the **original** form or version of the creator, maker or progenitor of the current cosmos, that is to say, if the latter are interested in understanding the form or version of the creator, maker or progenitor of the current cosmos before this creator, maker or progenitor of the current cosmos became the current ubiquitous, infinite and the *3-D or three-dimensional* **field of consciousness** called **cosmic space,** through the process of **expansion, distention, dilation** or **inflation** of its original or primordial self, namely its **dimensionless self,** in order to **spatially** accommodate the current **3-D** or **three-dimensional** physical cosmos inside itself.

Incidentally, the current **3-D** or **three-dimensional**

physical cosmos is nothing but the **condensed, compressed** or **compacted** form or version of a segment of the current ubiquitous, infinite and the *3-D or three-dimensional* **field of consciousness** called **cosmic space.**

Now let's pick up again the thread where it was left off earlier with regards to the above mentioned **vindu, point, confluence or sangam** of the human **sleep state** on one hand and the human **fully awake** or **wide-awake state** on the other.

At this juncture, it is imperative on one's part to point out once again that all the **human consciousnesses** at the **vindu, point, confluence** or **sangam** of the human **sleep state** on one hand and the human **wide-awake** or **fully awake state** on the other are, without fail, **aware** or **conscious** of themselves as an **aware** or **conscious being,** in the manner they were aware or conscious a little while earlier during their **REM sleep** or **dream sleep state.**

However, they are not aware of the presence of any **dream sleep state cosmos** inside their **consciousnesses** of which they were all aware or conscious a little while earlier during their **REM sleep** or **dream sleep states.**

By stating what one has stated above, one means that the **saangyic** or **brahmanic** human consciousnesses that exist at the **vindu, point, confluence** or **sangam** of the human **sleep state** on one hand and the human **fully awake** or **wide-awake state** on the other, are aware of themselves as **conscious beings** in the manner they were aware of themselves a little while earlier when they existed in their another form or version namely the form or version called **dream sleep state** human consciousnesses. However, as **dream sleep state** human consciousnesses during their **REM sleep** or **dream sleep**, they all were aware of the presence the **dream sleep state cosmoses** inside their **consciousnesses** which is not the case now because now they exist in their another form or incarnation called the **saangyic** or **brahmanic** human consciousnesses at this **vindu, point, confluence** or **sangam** of the human **sleep state** on one hand and the human **wide-awake** or **fully awake state** on the other.

Thus, the **saangyic** or **brahmanic** human consciousnesses that exists at this **vindu, point, confluence** or **sangam** of the human **sleep state** on one hand and the human **wide-awake** or **fully awake state** on the other are not aware of any kind of **dream sleep state cosmoses** inside their **consciousnesses** despite the fact that they are aware of themselves as **aware** or **conscious beings.** Hence, the **saangyic** or **brahmanic** human consciousnesses extant or present at this **vindu, point,**

confluence or **sangam** of the human **sleep state** on one hand and the human **wide-awake** or **fully awake state** on the other is a **dimensionless consciousness** in the manner creator, maker or progenitor of the current cosmos was in its **original** or **primordial** form, version or incarnation.

However, despite the possession on their part of this very **distinctive feature** or **attribute** that, they all without fail, are **aware** or **conscious** of themselves as an **aware** or **conscious beings**, the **saangyic** or **brahmanic** human consciousnesses that exists at the **vindu, point, confluence** or **sangam** of the human **sleep state** on one hand and the human **fully awake** or **wide-awake state** on the other, all are very **surprisingly** not aware of the existence of their **physical body** of which they all were aware or conscious very patently, palpably or perceptively when they all were in their **fully awake state** before going to sleep the previous night and of which they all will again become aware or conscious very patently, palpably or perceptively when they all will cross over to the other side namely the side called the **fully awake state,** following their very extraordinary **pause** for a very brief moment at this incredible twilight **vindu** or **point** or, **confluence** or **sangam.**

Thus, at this incredible twilight **bindu** or **point** or, twilight **confluence** or **sangam,** all **human consciousnesses** exist, albeit for a very brief moment, as an incredible **bodiless** and **dimensionless consciousness** in the manner creator, maker or progenitor of the current cosmos existed 13.7 billion light years ago, prior to the creation on its part of the latter namely the current cosmos.

The **bodiless** and **dimensionless** form or version of creator, maker or progenitor of the current cosmos was its **original form or version.** In this form or version it existed 13.7 billion light years ago, prior to the commencement on its part of its extraordinary act of creation called the current cosmos, through the process of **expansion, distention, dilation** or **inflation** of itself i.e. through the process of **expansion, distention, dilation or inflation** of its **dimensionless self** or **original self,** in order to give rise to the current **3-D** or **three-dimensional** plus ubiquitous and the infinite **field of consciousness** called the **cosmic space.**

The creator, maker or progenitor of the current cosmos had to create the current **3-D** or **three-dimensional** cosmic space first through the process of **expansion, distention, dilation or inflation** of its **dimensionless self** or of its **original self** because without the prior creation on its part of the current **3-D** or **three-dimensional** cosmic space, it would have been possible for it to **spatially** accommodate its **subsequent creation** namely the current **3-D** or **three-**

dimensional physical cosmos.

Just to remind oneself, the **bodiless** & **dimensionless** form or version or the **original** form or version of the creator, maker or progenitor of the current cosmos is also called its **saangyic, brahmanic** or **supra-transcendental** form or version.

To recap.

The current **cosmic space** is nothing but the **3-D** or **three-dimensional** form or version of creator, maker or progenitor of the current cosmos inside which the current **3-D** or **three-dimensional physical cosmos** is floating, wafting or levitating plus whirling, twirling or spiralling non-stop from the time of its inception some 13.7 billion light years ago and which also marked the inception, inauguration, birth, beginning dawn or debut of time itself.

Secondly, at the above described twilight **bindu** or **point** or, if it is preferred, at the above described twilight **confluence** or **sangam** of **human sleep state** on one hand and **human wakeful state** on the other, all **human consciousnesses** exist, albeit for a very brief moment, as an incredible **bodiless** and **dimensionless consciousness** in the manner creator, maker or progenitor of the current cosmos existed 13.7 billion light years ago, prior to the creation on its part of the latter namely the current cosmos.

Thirdly, at the above described twilight **bindu** or **point** or, if it is preferred, at the above described twilight **confluence** or **sangam** of **human sleep state** on one hand and **human wakeful state** on the other all human consciousnesses, despite being **aware of themselves as a conscious being,** are not aware of the **wakeful state** cosmos nor are aware of their physical body of which they were aware in their fully awake or wide-awake state.

Fourthly, at this twilight **bindu** or **point** or, at this twilight **confluence** or **sangam** of human **sleep state** on one hand and human **fully awake** or **wide-awake state** on the other, human consciousnesses are not aware of the presence of any kind of **dream sleep state cosmos** inside themselves in the manner human consciousnesses in their **dream sleep state** were aware of the presence of a **dream sleep state cosmos** inside themselves.

At this juncture, it will not be inappropriate to remind oneself once again that at the twilight **bindu** or **point** or, the twilight **confluence** or **sangam** of human **sleep state** on one hand and human **fully awake** or **wide-awake state** on the other, all human consciousnesses exist as an incredible **bodiless** and **dimensionless consciousness** in the manner creator, maker or progenitor of the current cosmos existed 13.7 billion light years ago when this creator was in its **original** form or version, that is to say, when this creator was in its **saangyic, brahmanic** or **supra-transcendental** form or version i.e. before this creator created the current cosmos consisting of the current **cosmic space,** the current **physical matter** of myriad forms and shapes plus the current **human consciousnesses** & all the other kinds of **consciousnesses** such as **animal consciousnesses.** This incredible or awe-inspiring and **the only one of its kind** or unique **timeless, bodiless** and **dimensionless** creator its current creation namely the current cosmos consisting of all the above- mentioned items, **out of its own consciousnessbal essence.**

What has been said up till now can be summarised as follows.

At the above described **vindu, point, confluence** or **sangam** of human **sleep state** on one hand and **human fully awake** or **wide-awake state** on the other, all human consciousnesses are **aware or conscious of themselves** (albeit for an extremely brief moment) **as an absolutely pure, formless, bodiless and dimensionless awareness** or **consciousness** and nothing else. Exactly this was condition of the creator, maker or progenitor of the current cosmos also, when it was extant or in existence at that particular **vindu** or **point** in its **original, fundamental, primal, primordial, inherent, innate, saangyic, brahmanic** or **supra-transcendental** form or version, 13.7 billion light years ago, before it metamorphosed itself into the current cosmos consisting of the current cosmic space, current physical matter of myriad forms and shapes, and all the current human consciousnesses plus all the other consciousnesses such as animal consciousnesses etc.

In other words, at this twilight **vindu, point, confluence** or **sangam** of human **sleep state** on one hand and **fully awake** or **wide-awake state state** on the other, **all** human consciousnesses are **absolutely alone** in the manner creator, maker or progenitor of the current cosmos was in its **original, saangyic, brahmanic, supra-transcendental** or **bodiless** and **dimensionless** form or version i.e. before it created the current cosmos consisting of the current **cosmic space,** current **physical matter** and the current **human consciousnesses** and all the other **consciousnesses** such as **animal consciousnesses.**

At this extraordinary twilight **vindu, point, confluence** or **sangam** of human **sleep state** on one hand and **fully awake** or **wide-awake state** on the other, the existent or extant human consciousnesses, even though they are aware or conscious of their own existence or, if it is preferred, are aware or conscious of themselves as an **aware or conscious being** but most surprisingly they neither aware or conscious of their **physical body** nor are aware or conscious of the **physical cosmos** of their **fully awake state** or **wide-awake state.**

It has already been stated above that these human consciousnesses which are extant or present at this extraordinary twilight **vindu, point, confluence** or **sangam** of human **sleep state** on one hand and **fully awake** or **wide-awake state** on the other, are neither aware of the existence of their own **physical body** of which they are aware in their **fully awake** or **wide-awake state** nor are they aware or conscious of the **physical cosmos** of which they are aware or conscious in their **fully awake** or **wide-awake state** but what need to be added that they are not aware or conscious of the presence or existence of any kind of **dream sleep state cosmos** inside their consciousnesses or awarenesses (in the manner they were aware or conscious during their **dream sleep state** or **REM sleep state**) at this extraordinary **bindu, point, confluence** or **sangam** of human **sleep state** on one hand and human **fully awake** or **wide-awake state** on the other.

Thus, at this extraordinary **vindu, point, confluence** or **sangam** of human **sleep state** on one hand and **fully awake** or **wide-awake state** on the other, these human consciousnesses or awarenesses exist or are extant as incredible **dimensionless consciousness** or **awarenesses** or, more to the point, as incredible **formless, bodiless** and **dimensionless consciousness** or **awarenesses** in the manner creator, maker or progenitor of the current cosmos existed or was extant 13.7 billion light years ago when it was in its **original, fundamental, primal, primordial, inherent, innate, saangyic, brahmanic** or **Supra-transcendental** form or version and was yet to transform or transmute itself into the present-day cosmos via the common-'o'-garden process of **daydreaming** on its part.

~*~*~*~*~

MATERIAL SCIENCE VS CONSCIOUSNESSBAL SCIENCE OF ADWAIT-VEDANTA

Apart from the presence of **cosmic space** in the current cosmos, there are two other ingredients which exist in the current cosmos namely the current **physical matter** of myriad shapes, sizes and attributes on one hand and the current **human consciousnesses** or **awarenesses** of myriad dispositions, proclivities or tendencies on the other, not to forget the myriad **consciousnesses** or **awarenesses** of other kinds, for example those of animals.

The **material science** is dependent for its **accuracy,**

authenticity or **veracity** as well as for its future **progress** or **advancement** on the **empirical** or the **posteriori knowledge,** that is to say, on the **knowledge** which has been garnered by the **material scientists** the world over from the **material** or the **physical world** following or after the **material** or the **physical sense-experiences** and **sense-observations** by them with the assistance of their **material** or **physical senses** or, if it is preferred, with the assistance of their **outer** or **external senses.**

Likewise, the **consciousnessbal science** of **Adwait-Vedanta** is dependent for its **accuracy, authenticity** or **veracity** as well as for its **progress** or **advancement** on the **empirical** or the **posteriori knowledge,** that is to say, on the **knowledge** which has been garnered by the **consciousnessbal scientists** the world over from the **consciousnessbal** or the **awarenessbal world** following or after the **consciousnessbal** or the **awarenessbal sense-experiences** and **sense-observations** by them with the assistance of their **consciousnessbal** or **awarenessbal senses** or, if it is preferred, with the assistance of their **inner** or the **internal senses.**

Both kinds of **sciences** i.e. the **material science** on one hand and the **consciousnessbal science** of **Adwait-Vedanta** on the other, add value to **human existence.**

The **material science** adds value to the **material**

dimension of human beings namely to their **material** or **physical bodies.**

The **consciousnessbal science** of **Adwait-Vedanta** on the other hand, adds value to the **consciousnessbal dimension** of human beings namely to their **consciousnesses** or **awarenesses.**

To test or evaluate plus confirm or affirm the **veracity, accuracy** or **authenticity** of the **material** or **physical truths** of the **world** of **material** or **physical science,** the **material scientists** use all kinds of **material** or **physical gadgets** or **vehicles,** for example **material** or **physical Large Hadron Collider, material** or **physical laboratories** of all kinds and the like.

Likewise, the **consciousnessbal scientists** use their own **consciousnesses** as their personal or private **consciousnessbal gadgets** or **vehicles** or, if one likes, as their personal or private **consciousnessbal Large Hadron Colliders** or **consciousnessbal laboratories** to verify and re-verify the **veracity** or **authenticity** of their **consciousnessbal experiences** and **observations.** They also urge other **consciousnessbal scientists** to do the same so that these **consciousnessbal experiences** and **observations** can be tested again and again **consciousnessbally** by the **consciousnessbal scientists** the world over. Through this technique, these **consciousnessbal experiences** and **observations** of **consciousnessbal scientists** have been authenticated or validated over and over throughout the world. On account of this, these **consciousnessbal experiences** and **observations** of the **consciousnessbal scientists** have become accepted as being **empirical** or **posteriori knowledge** or **truths** of the **consciousnessbal variety** amongst the **consciousnessbal scientists** the world over.

What has been said above can be put in another way.

The manner in which , the above mentioned **consciousnessbal experiences** and **observations** of the **consciousnessbal science** of **Adwait-Vedanta** have been tested again and again by the **consciousnessbal scientists** the world over, it has earned them the right to be accepted universally as being one hundred percent genuine or authentic, **empirical** or **posteriori knowledge** or **truths** of the **consciousnessbal variety** which are on par with or equal to, if not superior to, the **empirical** or **posteriori knowledge** or **truths** of **material** or **physical variety** as far as their importance with regards to mankind's **consciousnessbal existence** is concerned , the **consciousnessbal knowledge** or **truths** whose veracity or authenticity have been authenticated or verified **consciousnessbally** again and again by the **consciousnessbal scientists** the world over, plus whose veracity or authenticity can be **consciousnessbally** re-checked or re-scrutinised by anyone at any time inside their own, personal, or private **consciousnessbal Large Hadron Colliders** or **consciousnessbal laboratories** namely their **consciousnesses** or **awarenesses.**

The **authenticity** of the **knowledge** or **truth,** presented to mankind by the **consciousnessbal science** of **Adwait-Vedanta** with regards to the **true nature** and the **true cause** of the current cosmos is based on the **experiential evidences** of the **consciousnessbal variety,** garnered by the **consciousnessbal scientists** the world over via the direct, firsthand, hands-on, face-to-face or empiric **observations** done by their **consciousnessbal senses** during their nightly **dream sleep states** on one hand, and via the direct, first hand, hands-on, face-to-face or empiric **experimental monitoring, scrutiny** or **surveillance** of the **consciousnessbal variety** undertaken by these **consciousnessbal scientists** with the assistance of their **consciousnessbal senses** during their **daydreaming-dexterity** or **deftness** in their **fully awake** or **wide-awake states** on the other.

In other words, the **authenticity** of the **truths** expounded by the **consciousnessbal science** of **Adwait-Vedanta** with regards to the **true nature** and the **true cause** of the current cosmos is deeply-rooted in factual, truthful or authentic, **experiential evidences** of **consciousnessbal variety,** garnered directly, first hand, hands-on, face-to-face or empirically by the **consciousnessbal senses** of the **consciousnessbal scientists** the world over during their nightly **dream sleep state's experiences** on one hand, and direct, first hand, hands-on, face-to-face or empiric, **experimental observations** and **investigations** of **consciousnessbal kind** undertaken by these **consciousnessbal scientists** inside their **consciousnesses** with the assistance of their **consciousnessbal senses** during their **daydreaming-dexterity** or **deftness** in their **fully awake** or **wide-awake states** on the other.

Hence the **knowledge, fact** or the **judgement** plus the **information, understanding** or the **realisation,** offered to mankind by the **consciousnessbal science** of **Adwait-Vedanta** with regards to the **true nature** and the **true cause** of the current cosmos are not based on reason or logic, nor on theory, hypothesis, or speculation. Neither are they based on postulation, supposition, deduction or inference. That is to say, they are not based on **priori** or **presumption.**

The knowledge put forward by the **consciousnessbal science** of **Adwait-Vedanta** with regards to the **true nature** and the **true cause** of the current cosmos are provable, verifiable, confirmable or testable by all **human consciousnesses** through their passive but nonetheless direct, firsthand, hands-on, face-to-face or empiric **consciousnessbal experiences** during their nightly **dream sleep states** on one hand, and through their deliberate or done on purpose direct, firsthand, hands-on, face-to-face or empiric **consciousnessbal experimentations, investigations** and **observations** during their **daydreaming-dexterity** or **deftness** in their **fully awake** or **wide-awake states** on the other.

In other words, the **knowledge** or the **truths** presented to mankind by the **consciousnessbal science** of **Adwait-Vedanta** with regards to the **true nature** and the **true cause** of the current cosmos are not conjectures or assumptions.

Instead, they are based on rock-solid, accurate or authentic, albeit passive **consciousnessbal experiences** of their nightly **dream sleep states** on one hand and hard-headed or unsentimental plus practical, pragmatic or businesslike, active **consciousnessbal experimentations, observations** and **investigations** performed via their **daydreaming-dexterity** or **deftness** during their **fully awake** or **wide-awake states** on the other.

Thus, the knowledge offered to mankind with regards to the **true nature** and the **true cause** of the current cosmos by the **consciousnessbal science** of **Adwait-Vedanta** is mankind's **consciousnessbal-senses** based or **awarenessbal-senses** based or, if it is preferred, is mankind's **inner-senses** based or **internal-senses** based.

In other words, the knowledge, conclusions or judgements put forward to mankind by the **consciousnessbal science** of **Adwait-Vedanta** with regards to the **true nature** and the **true cause** of the **current cosmos** are acquired directly , firsthand, hands-on, face-to-face or empirically by the **consciousnesses** or the **awarenesses** of the **consciousnessbal scientists** via their **consciousnessbal-senses** or **awarenessbal-senses** or, if it is preferred, via their **inner-senses** or **internal-senses** through the passive **consciousnessbal** or **awarenessbal experiences** of their nightly **dream sleep states** on one hand and via their active **consciousnessbal** or **awarenessbal experimentations, observations** and **investigations** during their **daydreaming-dexterity** or **deftness** in their **fully awake** or **wide-awake states** on the other.

To repeat.

The knowledge, conclusions or judgements put forward by the **consciousnessbal   science** of **Adwait-Vedanta** with regards to the **true nature** and the **true cause** of the **current cosmos** are provable, verifiable, confirmable or testable by all **human consciousnesses** through their passive but nonetheless direct, first hand, hands-on, face-to-face or empiric **consciousnessbal experiences** via their **consciousnessbal-senses** or **awarenessbal-senses** or, if it is preferred, via their **inner-senses** or **internal-senses** during their nightly **dream sleep**

states on one hand and through their active direct, firsthand, hands-on, face-to-face or empiric **consciousnessbal experimentations, observations** and **investigations** via their **consciousnessbal-senses** or **awarenessbal-senses** or, if it is preferred, via their **inner-senses** or **internal-senses** during their **daydreaming-dexterity** or **deftness** in their **fully awake** or **wide-awake states** on the other.

These conclusions or facts with regards to the **true nature** and the **true cause** of the **current cosmos** which have been presented to mankind by the **consciousnessbal science** of **Adwait-Vedanta** are therefore not conjectures, assumptions or faith-based pronouncements.

Instead, they are rock-solid **empirical** or **posteriori truths** which have been garnered **consciousnessbally** by the **consciousnessbal scientists** the world over directly, firsthand, hands-on, face-to-face or empirically via their **consciousnessbal senses** or **awarenessbal senses** or, if it is preferred, via their **inner senses** or **internal senses** during their nightly **dream sleep state's experiences** on one hand and through the medium of the **consciousnessbal experimentations, observations** and **investigations,** undertaken by them during their **daydreaming-dexterity** or **deftness** in their **fully awake** or **wide-awake states** on the other. These facts can be tested **consciousnessbally** directly, firsthand, hands-on, face-to-face or empirically by any **human consciousness** in the world, at any time.

What has been said above can be further explained in the following manner.

The knowledge put forward to the world by the **consciousnessbal science** of **Adwait-Vedanta** with regards to the **true nature** and the **true cause** of the current cosmos is mankind's **consciousnessbal-senses**-based or **awarenessbal-senses**-based knowledge or, if it preferred, is mankind's **inner-senses**-based or **internal-senses**-based knowledge and is not logic-based or reasoning-based knowledge.

In other words, knowledge presented to the world by the **consciousnessbal science** of **Adwait-Vedanta** with regards to the **true nature** and the **true cause** of the current cosmos is acquired through the medium of mankind's **consciousnessbal-senses** or the **awarenessbal-senses** or, if it is preferred, is acquired through the medium of mankind's **inner-senses** or the **internal-senses**, via the passive **consciousnessbal** or **awarenessbal experiences** of their nightly **dream sleep states** on one hand and via their active **experimentations, observations** and **investigations** during their **daydreaming-dexterity** or **deftness** in their **fully awake** or **wide-awake states** on the other.

As a result, the **consciousnessbal experiences** garnered and the **consciousnessbal observations** made by the **consciousnessbal scientists** the world over with the assistance of their **consciousnessbal senses** or **awarenessbal senses** or if one prefers, with the assistance of their **inner senses** or **internal senses** plus the **pronouncements** communicated to mankind by these **consciousnessbal scientists** through the **consciousnessbal science** of **Adwait-Vedanta** are accepted as being one hundred percent **empirically genuine** by the **consciousnessbal scientists** the world over.

~*~*~*~*~

CONSCIOUSNESSBAL SCIENCE OF ADWAIT-VEDANTA PROVIDES A POSTERIORI OR EMPIRICAL KNOWLEDGE AND NOT A PRIORI OR PRESUMPTIVE KNOWLEDGE

The knowledge provided to mankind about the true cause and the true nature of the current cosmos, by the **consciousnessbal science** of **Advait-Vedanta** is a **posteriori** or **empirical knowledge** and not a **priory** or **presumptive knowledge.**

When one says that the knowledge imparted to mankind about the true cause and the true nature of the current cosmos, by the **consciousnessbal** science of **Adwait-Vedanta,** is a **posteriori** or **empirical knowledge** and not a **priori** or **presumptive knowledge,** one means that this **Adwait-Vedantic scientific knowledge** of **consciousnessbal variety** is not based on logic or reason. Nor is it based on theory, hypothesis, speculation, postulation, supposition, deduction, inference, opinion, view point or belief system.

Instead, this **Adwait-Vedantic scientific knowledge** of **consciousnessbal variety** is based on rock solid or up-front experiential or existential as well as hard-boiled or tough experimental or investigational evidences of **consciousnessbal** or **awarenessbal** variety. In other words, it is based on thoroughgoing or hard-headed factual or practical as well as absolute or out-and-out observed or witnessed evidences of **consciousnessbal** or **awarenessbal** variety.

The facts made available to mankind by the **consciousnessbal science** of **Adwait-Vedanta** are provable, verifiable, confirmable or testable by all human beings through their personal, **consciousnessbal experiences** or **practical contacts** of their **dream sleep states** each night on one hand and through their personal **consciousnessbal experiments** or **investigations** via the medium of their natural, **consciousnessbal daydreaming-dexterity** or **deftness** in their **fully awake** or **wide-awake states** on the other.

Thus, the knowledge afforded to mankind by the **consciousnessbal science** of **Adwait-Vedanta** is not a subjective conjecture, guess, speculation, surmise, belief, assumption, theory, hypothesis, postulation or supposition.

The **consciousnessbal experiences** or **personal contacts** provided to **human consciousnesses** passively during their **dream sleep states** each night on one hand, as well as the **consciousnessbal experimental** or **investigational observations** or **scrutiny,** secured by **human consciousnesses through** the active **experimentations** or **investigations** on their part, with the aid of their **consciousnessbal daydreaming-dexterity** or **deftness** on the other, are all **consciousnessbal-senses** based or **awarenessbal-senses** based or if it is preferred, are all **inner-senses** based or all **internal-senses** based and not based on reasoning, logic, theory, hypothesis, speculation, postulation, supposition, deduction, inference, opinion, view point or belief system.

In other words, knowledge, conclusions or judgements offered to mankind by the **consciousnessbal science** of **Adwait-Vedanta** have been acquired by means of **consciousnessbal** or **awarenessbal senses** of human beings directly, personally, in person, face to face or at first-hand or, if it is preferred, has been acquired by means of **inner** or **internal senses** of human beings directly, personally, in person, face to face or at first-hand, through the **consciousnessbal** or **awarenessbal** passive observations, monitoring, viewings or scrutiny during their **dream sleep states** each night on one hand and through the **consciousnessbal** or **awarenessbal** active experimentations or investigations on their part during their **daydreaming-dexterity** or **deftness** in fully awake or wide-awake state on the other hand.

Annex

A small note about the five outer, external, physical or material senses of mankind and five inner, internal, consciousnessbal or awarenessbal senses of mankind.

Just as human beings have their five physical, material, outer or external senses namely those of seeing (eyes), hearing (ears), smell (nose), taste (tongue) and touch (skin) to perceive and exprcience the current **fully awake** or **wide-awake state cosmos,** likewise they also possess or rather have been endowed on purpose or deliberately by their creator, maker or progenitor, their five consciousnessbal, awarenessbal, inner or internal senses of seeing, hearing, smell, taste and touch which operate, work, function or are in action and thus assume immense importance during mankind's passive but nonetheless direct, personal, in person, face to face or at first-hand, consciousnessbal or awarenessbal observations, monitoring, viewings or scrutiny of their **dream sleep states** each night on one hand and during their active and at the same time direct, personal, in person, face to face or at first-hand, consciousnessbal or awarenessbal **daydreaming** experiments or investigations through the use on their part of their **daydreaming-dexterity** or **deftness** in their fully **fully-awake** or **wide-awake state** on the other hand.

~*~*~*~*~

OCEAN OF CONSCIOUSNESS VS OCEAN OF WATER

Just as the **ocean of water** on the surface of our earth is much bigger than the **islands of ice,** which are currently seen floating   or wafting in this **ocean of water,** at the north and south poles of the earth, likewise the **ocean of consciousness** aka **cosmic space** in the current cosmos is much bigger than the **islands of physical matter,** for example, planets, stars and galaxies etc. which are currently seen floating or wafting   all over the place in this **ocean of consciousness** aka **cosmic space.**

Just as the **islands of ice,** seen floating or wafting   in the **ocean of water** are nothing but the **solidified** form or version of a segment of this **ocean of water,** likewise the **islands of physical matter,** for instance, moons, planets, stars and galaxies etc., seen floating, wafting or levitating plus whirling, twirling or spiralling in this **ocean of consciousness** aka **cosmic space,** are nothing but the **solidified** form or version of a segment of this **ocean of consciousness** aka **cosmic space.**

Incidentally, the current ubiquitous and the infinite **ocean of consciousness** aka **cosmic space** is nothing but the **expanded, distended, dilated** or **inflated** form or version of the **original** or **primordial** form or version of creator, maker or progenitor of the current cosmos.

And the **original** or **primordial** form or version of creator, maker or progenitor of the current cosmos is described as the incredible or awe-inspiring and **the only one of its kind** or unique **timeless, bodiless** and **dimensionless consciousness** of infinite intelligence and emotion

SUPRA-TRANSCENDENTAL ANUBHUTI OR EXPERIENCE

The term **supra-transcendental** anubhuti or experience refers to the anubhuti or experience which is very like, akin, near, or close to the anubhuti or experience of the **original** or **primordial** form or version or, if it is preferred, close to the **saangyic** or **brahmanic** form or version of creator, maker or progenitor of the current cosmos.

The **supra-transcendental, original, primordial, saangyic,** or **brahmanic** form or version of creator, maker or progenitor of the current cosmos is described as the incredible or awe-inspiring and **the only one of its kind** or unique, **timeless, bodiless** and

dimensionless consciousness or **awareness** of infinite intelligence and emotion.

In its above described, **supra-transcendental, saangyic, brahmanic, original** or **primordial** form or version, this creator, maker or the progenitor of the current cosmos existed some 13.7 billion light years ago.

However, currently this creator, maker or the progenitor of the current cosmos exists as the amazing, **3-D** or **three-dimensional,** ubiquitous and the infinite **field of consciousness** or, if one prefers, as the amazing, **3-D** or **three-dimensional,** ubiquitous and the infinite **ocean of consciousness** called, **cosmic space.**

This amazing, **3-D** or **three-dimensional,** ubiquitous and the infinite **field of consciousness** or, this amazing, **3-D** or **three-dimensional,** ubiquitous and the infinite **ocean of consciousness** called **cosmic space** is expanded, distended, dilated or inflated form or version of **timeless, bodiless** and **dimensionless consciousness** of creator, maker or the progenitor of the current cosmos.

In this amazing, **3-D** or **three-dimensional,** ubiquitous and the infinite **field of consciousness** or, in this amazing, **3-D** or **three-dimensional,** ubiquitous and the infinite **ocean of consciousness** aka **cosmic space,** the current, **3-D** or **three-dimensional** *physical* *cosmos* is floating, wafting or levitating plus whirling, twirling or spiralling non-stop from the time of its inception, as tiny islands of **solidified** form or version of a segment of this amazing, **3-D** or **three-dimensional**, ubiquitous and the infinite **field of consciousness** or **ocean of consciousness** aka **cosmic space.**

Incidentally, the point of inception of the current **physical cosmos** also heralded or signalled the beginning or the birth of the current **time** which, like the current **physical cosmos** is also about 13.7 billion light years old.

When **human consciousnesses** begin to ascend each morning from their **sleep state** into their **fully awake** or **wide-awake state**, they meet or encounter an amazing **vindu** or **point** or, if it is preferred, an amazing **sandhi, sangam** or **junction** between their **sleep-state** on one hand and their **fully awake** or **wide-awake state** on the other.

The duration of this amazing **vindu, point, sandhi, sangam** or **junction** between mankind's **sleep-state** on one hand and their **fully awake** or **wide-awake state** on the other varies quite a great deal from one person to another, but generally speaking, it is of an extremely transient duration. It is a kind of **twilight zone** that exists for a fleeing moment between mankind's **sleep state** on one hand and their **fully** awake or **wide-awake state** on the other.

This amazing **vindu, point, sandhi, sangam** or **junction** between mankind's **sleep-state** on one hand and their **fully awake** or **wide-awake state** on the other is called **saangyic** or **brahmanic vindu, point, sandhi, sangam** or **junction.**

This amazing, **saangyic** or **brahmanic vindu, point, sandhi, sangam** or **junction** between mankind's **sleep-state** on one hand and their **fully awake** or **wide-awake state** on the other, as said before, is of an extremely short duration.

The above mentioned **saangyic** or **brahmanic vindu, point, sandhi, sangam** or **junction** of an extremely brief duration, between mankind's **sleep-state** on one hand and their **fully awake** or **wide-awake state** on the other, can be explained as a very brief period of **ambiguity** during which one is neither in one's **sleep state** nor in one's **fully awake** or **wide-awake state.**

Alternatively, this **saangyic** or **brahmanic vindu, point, sandhi, sangam** or **junction** between mankind's **sleep-state** on one hand and their **fully awake** or **wide-awake state** on the other can be narrated as an amazing but an extremely brief period of **half-sleep** and **half-awake state.**

The **saangyic** or **brahmanic vindu, point, sandhi,**

sangam or **junction** has a great significance in the realm of the **consciousnessbal science** of **Adwait-Vedanta**. It has therefore been accorded a **special tag** called the " **saangyic anubhuti-vindu** " or the " **brahmanic anubhuti-vindu**".

As said before, the period for which this " **saangyic anubhuti-vindu** " or this " **brahmanic anubhuti-vindu** " lasts, endures or stays around at this **sandhi, sangam** or **junction** of mankind's **sleep state** on one hand and their **fully awake** or **wide-awake state** on the other, is very brief but its significance is immense.

The significance of this " **saangyic anubhuti-vindu** " or this "**brahmanic anubhuti-vindu** " is immense because at this **vindu, point, sandhi, sangam** or **junction, human consciousnesses** encounter for a very brief period the **direct** or the **first-hand anubhuti** or **experience** of **bodiless-ness** and **dimensionless-ness.**

In other words, at this **vindu, point, sandhi, sangam** or **junction** each morning, **human consciousnesses** for a very brief moment become **bodiless** and **dimensionless consciousnesses** or **supra-transcendental, saangyic** or **brahmanic consciousnesses.**

The **bodiless** and **dimensionless** *human consciousnesses*, that exist at this incredible **supra-**

transcendental, saangyic or **brahmanic vindu, point, sandhi, sangam,** or **junction,** are neither in their **sleep state** nor in their **fully awake** or **wide-awake state.**

At this **vindu, point, sandhi, sangam** or **junction,** they are unaware of their **physical body** of which they were aware in their **fully awake** or **wide-awake state.** Furthermore, they are also unaware of the **fully awake** or **wide-awake state cosmos.**

However, at this **vindu, point, sandhi, sangam** or **junction** they are **aware** or **conscious** of themselves as an **aware being.** That is to say, at this **vindu, point, sandhi, sangam** or **junction** they are **aware** or **conscious** of their **existence** or **being** as an **aware** or **conscious being.** Hence, at this **vindu, point, sandhi, sangam** or **junction,** *human consciousnesses* or *awarenesses* for a very brief period become **bodiless and dimensionless** *consciousnesses* or *awarenesses.*

To sum up.

At the above narrated **supra-transcendental, saangyic** or **brahmanic vindu, point, sandhi, sangam** or **junction, all** *human consciousnesses,* for a very brief period, become **bodiless** and **dimensionless consciousnesses.**

At this **vindu, point, sandhi, sangam** or **junction,** *human consciousnesses* do not contain or possess inside them, any kind of **dream-stuff composed**

cosmos, that is to say, any kind of **dream-stuff composed cosmos** either of the **daydreaming variety** or of the **dream sleep variety.**

They thus, at this **vindu, point, sandhi, sangam** or **junction** do not contain or possess any kind of **3-D** or **three-dimensional** *space* or, more to the point, any kind of **3-D** or **three-dimensional** *cosmic space* inside them because at this **vindu, point, sandhi, sangam** or **junction** there is no need for them to hold or contain such a **space** or, more to the point, such a **cosmic space** inside them, since at this **vindu, point, sandhi, sangam** or **junction,** they are neither **daydreaming** nor are they in their **dream sleep state.**

They thus, at this **vindu, point, sandhi, sangam** or **junction,** are an incredible **bodiless** and **dimensionless consciousness.**

What has been said above can be put in another way.

At the above mentioned **vindu, point, sandhi, sangam** or **junction** of their **sleep-state** on one hand and their **fully awake** or **wide-awake state** on the other, all *human consciousnesses* exist or abide as **pristine** or **pure, bodiless** and **dimensionless consciousnesses,** in the manner creator, maker or the progenitor of the current cosmos existed some 13.7 billion light years ago when this creator, maker or the progenitor of the current cosmos was in its **original, primordial, saangyic, brahmanic** or **supra-transcendental** form or version.

As said before, the incredible creator, maker or progenitor of the current cosmos existed in its **original, primordial, saangyic, brahmanic** or **supra-transcendental** form or version some 13.7 billion light years ago.

This **original, primordial, saangyic, brahmanic** or **supra-transcendental** form or version of creator, maker or the progenitor of the current cosmos is described as the incredible or awe-inspiring, **the only one of its kind** or unique, **timeless, bodiless** and **dimensionless consciousness** or **awareness** of infinite intelligence and emotion.

~*~*~*~*~

SUPRA TRANSCENDENTAL, TRANSCENDENTAL,& INFRA-TRANSCENDENTAL CONSCIOUSNESSES- 1

A. Supra-Transcendental Consciousness

The **original, primordial, saangyic,** or **brahmanic** form or version of the creator, maker or progenitor of the current cosmos - which is described as being the incredible or awe-inspiring and **the only one of its kind** or unique, **timeless, bodiless** and **dimensionless consciousness** or **awareness** of infinite intelligence and emotion and which existed in its **original** form or version some 13.7 billion light years ago but which currently exists as the amazing, ubiquitous and the infinite, **3-D** or **three-dimensional field of consciousness** or, **ocean of consciousness** called **cosmic space -** has another designation namely the **Supra-Transcendental Consciousness.**

B. Transcendental and Infra-transcendental Consciousnesses

In contrast to the incredible or awe-inspiring and **the only one of its kind** or unique **timeless, bodiless** and **dimensionless Supra-Transcendental Consciousness** of infinite intelligence and emotion, (which is eternally singular, solitary or one and only), there are currently many billions of **human consciousnesses** in existence on the present-day planet earth.

The two entirely different adjectives, namely **transcendental** and **infra-transcendental,** are used whilst describing these **human consciousnesses.**

The use of two different adjectives instead of one, in relation to current **human consciousnesses,** is necessitated because one needs to distinguish them or, better still, one needs to place them into two distinct categories , on the basis of their knowledge with regards to the true nature of the current **cosmic space** on one hand and concerning the true nature of the current **human consciousnesses** plus the true nature of the current **physical matter** on the other, both of which are currently in existence in the present-day cosmos.

Those **human consciousnesses** who are aware unequivocally, categorically, indubitably or, without any ifs and buts of the true nature of all the three, fundamental **constituent-items** of the current cosmos namely, the **cosmic space**, the **human consciousnesses** and the **physical matter,** on the line described below, are called **transcendental human consciousnesses** and those who do not, are called **infra-transcendental human consciousnesses.**

Below one has described the true nature of all the three, fundamental **constituent-items** of the current cosmos namely, the **cosmic space**, the **human consciousnesses** and the **physical matter.**

COSMIC SPACE

The current **cosmic space** is not - as most **human consciousnesses** think it is - some dead to the world, inert, lacking perception, lacking feeling, lacking understanding, lacking consciousness, insentient or insensate thing which just happens to be there by serendipity, in order to spatially accommodate the **3-D** or the **three-dimensional physical matter** of the current cosmos, inside its **3-D** or **three-dimensional** space. Instead, the truth is that this **cosmic space** is an amazing, ubiquitous and infinite, **3-D** or **three-dimensional field of consciousness** or, **ocean of**

consciousness, created by the creator, maker or progenitor of the current cosmos through the process of **expansion, distention, dilation** or **inflation** of itself or, more to the point, through the process of **expansion, distention, dilation** or **inflation** of its **original** form or version. This **original** form or version of the creator, maker or progenitor of the current cosmos is also called this creator's **primordial, saangyic, brahmanic** or **suprs-transcendental** form or version.

The creator, maker or progenitor of the current cosmos existed in its **original** form or version some 13.7 billion light years ago. As indicated above, now it does not exist in its **original** form or version. Instead, now it exists as the amazing, ubiquitous and infinite, **3-D** or **three-dimensional field of consciousness** or, **ocean of consciousness** which is ignorantly called **cosmic space** by the vast majority of the current **human consciousnesses.**

The **original** form or version of the creator, maker or progenitor of the current cosmos is described as the incredible or awe-inspiring and **the only one of its kind** or unique, **timeless, bodiless** and **dimensionless consciousness** or **awareness** of infinite intelligence and emotion.

HUMAN CONSCIOUSNESSES

Each and every current **human consciousness** which

is in existence on the present-day planet earth, is a pure or pristine form or version of a segment of the amazing, ubiquitous and infinite **3-D** or **three-dimensional field of consciousness** or, **ocean of consciousness** which is ignorantly called **cosmic space** by the vast majority of the current **human consciousnesses.**

PHYSICAL MATTER

Last but not least, the **physical matter** which is in existence in the present-day **cosmos** and constitutes the **physical bodies** of human beings, in addition to constituting the remainder of the current **physical cosmos,** is nothing but merely a **condensed, compressed** or **compacted** form or version of a segment of the amazing, ubiquitous and infinite **3-D** or **three-dimensional field of consciousness** or, **ocean of consciousness,** ignorantly called **cosmic space** by the vast majority of the current **human consciousnesses.**

~*~*~*~*~

SUPRA-TRANSCENDENTAL, TRANSCENDENTAL, AND INFRA-TRANSCENDENTAL CONSCIOUSNESSES- 2

On account of the lack of adequate and proper application of thought, the vast majority of **human consciousnesses** tie their identity very firmly to their **time-bound, temporal,** or **mortal 3-D** or **three-dimensional** and thus **space-occupying physical body.**

This incorrect line of thinking on their part makes them believe that they are the **time-bound, temporal,** or **mortal, 3-D** or **three-dimensional** and thus **space-occupying, physical bodies** rather than the

timeless, dimensionless, and thus **non-space-occupying consciousnesses** which in reality they are.

To repeat.

The vast majority of **human consciousnesses** which exist on the present-day planet earth, cling very firmly to their delusion or wrong notion that their true identity consists of their **time-bound, temporal,** or **mortal, 3-D** or **three-dimensional** and thus **space-occupying, physical body.**

They do not know that they are not the **time-bound, temporal,** or **mortal, 3-D** or **three-dimensional** and thus **space-occupying physical body.** Instead, they are the **immortal, eternal,** or **timeless,** plus **dimensionless,** and thus **non-space-occupying consciousness.**

They also do not know that their **time-bound, temporal,** or **mortal, 3-D** or **three-dimensional,** and thus **space-occupying physical body** is merely their temporary house or residence inside which they had been housed or lodged by their **fountainhead** or **source** namely, the amazing, ubiquitous, and infinite, **3-D** or **three-dimensional, field of consciousness,** or **ocean of consciousness** called **cosmic space.**

The amazing, ubiquitous, and infinite, **3-D** or **three-dimensional, field of consciousness,** or **ocean of** consciousness, called **cosmic space**, has temporarily housed or lodged the **immortal, eternal,** or **timeless**, plus **dimensionless**, and thus, **non-space-occupying, human consciousnesses**, inside their own, personal or individual, **time-bound, temporal,** or **mortal, 3-D** or **three-dimensional**, and thus, **space-occupying, physical bodies**, in order purely to enact the brilliant, **cosmic play**, or **cosmic drama**, or if it is preferred, in order purely to stage the brilliant, **cosmic dance** of ' '**I** and **my'** ' **you** and **yours** ' , ' **they** and **theirs** ' , ' **this** and **that** ', **good** and **bad, saint** and **sinner, saviour** and **slayer, beauty** and **beast, genius** and **jerk, black** and **brown, white** and **yellow, tall** and **small,** so on and so forth. This brilliant **cosmic play** or **cosmic drama** or, this brilliant **cosmic dance** is none other than the very complex and enigmatic entity called the present **physical** or **material cosmos.**

What has been said above can be put in another way.

The amazing, ubiquitous, and infinite, **3-D** or **three-dimensional, field of consciousness,** or **ocean of consciousness,** called **cosmic space,** has temporarily housed or lodged the **immortal, eternal,** or **timeless**, plus **dimensionless**, and thus, **non-space-occupying, human consciousnesses,** inside their own, personal or individual, **time-bound, temporal,** or **mortal,** plus **3-D** or **three-dimensional,** and thus, **space-occupying, physical bodies,** in order purely to create **plurality, diversity, variety, multiplicity,** or **heterogeneity,** inside its inherently, nondescript or featureless, **consciousness,** with the sole purpose of beguiling, amusing, entertaining, or regaling itself and nothing else. And this **primogenitor-consciousness,** who is the **fountainhead** or **source** of the **immortal, eternal,** or **timeless,** plus **dimensionless,** and thus, **non-space-occupying, human consciousnesses,** on one hand, and their private, personal, or individual, **time-bound, temporal,** or **mortal, 3-D** or **three-dimensional,** and thus, **space-occupying, physical bodies,** on the other, nay, who is the **fountainhead,** or **source,** of the entire current cosmos, is none other than the entity called the amazing, ubiquitous, and infinite, **3-D** or **three-dimensional, field of consciousness,** or **ocean of consciousness,** called **cosmic space,** which, in turn, is nothing but the **expanded, distended, dilated,** or **inflated,** form or version of the **original** form or version of the creator, maker, or progenitor, of the current cosmos.

The **original** form or version of the creator, maker, or progenitor, of the current cosmos existed some 13.7 billion light years ago. It does not exist now in that form or version. Presently, it exists as the amazing, ubiquitous and infinite, **3-D** or **three-dimensional field of consciousness** or, **ocean of consciousness** called **cosmic space.**

The **original** form or version of the creator, maker or progenitor of the current cosmos, which existed some 13.7 billion light years ago, is described as the incredible or awe-inspiring and **the only one of its kind** or unique, **timeless, bodiless,** and **dimensionless consciousness** or **awareness** of infinite intelligence, imagination, and emotion.

To recap.

The vast majority of **immortal, eternal,** or **timeless** plus **dimensionless** and thus, **non-space-occupying, human consciousnesses,** which exist on the present-day planet earth, cling very firmly to their delusion or wrong notion that their true identity consists of their **time-bound, temporal,** or **mortal,** plus **3-D** or **three-dimensional** and thus **space-occupying physical body.**

However, the reality is quite different.

This reality is that the **true identity** of the present-day, **immortal, eternal,** or **timeless** plus **dimensionless** and thus, **non-space-occupying, human consciousnesses** does not consists of their **time-bound, temporal,** or **mortal** plus **3-D** or **three-dimensional** and thus **space-occupying physical bodies.** Instead, their true identity consists of their **immortal, eternal,** or **timeless,** plus **dimensionless** and thus **non-space-occupying consciousnesses** in the manner of their **fountainhead** or **source** namely, the creator, maker or progenitor of the current cosmos.

The virtuoso creator, maker or progenitor of the current cosmos possesses an innate power or ability to exist in two forms or versions, one at a time, as per the vagary or caprice of its mood, whim, or fancy. These two forms or versions of this virtuoso creator, maker or progenitor of the current cosmos has already been described above but it may not be inappropriate to recount them once again.

Firstly, this brilliant creator, maker or progenitor of the current cosmos existed in its **original** form or version some 13.7 billion light years ago only. And so, it does not exist now in that form or version. The **original** form or version of the creator, maker or progenitor of the current cosmos is described as the incredible or awe-inspiring and **the only one of its kind** or unique, **timeless, bodiless, and dimensionless, consciousness** or **awareness** of infinite intelligence, imagination, and emotion.

Presently, this gifted creator, maker or progenitor of the current cosmos exists as the amazing, ubiquitous and infinite, **3-D** or **three-dimensional field of consciousness** or, **ocean of consciousness** called **cosmic space.**

Hence, the entity currently called the **cosmic space** is nothing but the **expanded, distended, dilated** or

inflated form or version of the **original** form or version of the creator, maker or progenitor of the current cosmos and the **original** form or version of the creator, maker or progenitor of the current cosmos is its **dimensionless** form or version. Therefore, the current **cosmic space** is nothing but the **3-D** or **three-dimensional** form or version of the **original, dimensionless,** form or version of the creator, maker or progenitor of the current cosmos

Now let's return to the topic which was earlier under our focus namely, what is the **true identity** of the **human consciousnesses?**

As said before, despite the fact that the vast majority of **human consciousnesses** cling very firmly to their delusion or wrong notion that their **identity** consists of their **time-bound, temporal,** or **mortal, 3-D or three-dimensional** and thus **space-occupying physical body,** the reality is that their **true identity** consists of the **immortal, eternal,** or **timeless,** plus **dimensionless and non-space-occupying consciousness.** Their **time-bound, temporal,** or **mortal** plus **3-D or three-dimensional** and thus **space-occupying physical bodies** are merely their temporary **houses** or **residences** in which they have been housed or lodged by their **fountainhead** or **source** namely, the amazing, ubiquitous and infinite, **3-D or three-dimensional field of consciousness** or, **ocean of consciousness** called **cosmic space** which, in turn, is nothing but the **expanded, distended, dilated** or

inflated form or version of the **original, dimensionless** form or version of the creator, maker or progenitor of the current cosmos.

Hence, even though the **human physical bodies,** inside which the current **human consciousnesses** have been temporarily housed or lodged by their **fountainhead** or **source,** are **time-bound, temporal,** or **mortal,** the **human consciousnesses** themselves are **timeless, eternal,** or **immortal** in the manner of their **fountainhead** or **source.** And this **fountainhead** or **source** of the **human consciousnesses** is the self-same or the very same, incredible or awe-inspiring, creator, maker or progenitor of the current cosmos who possesses infinite intelligence, imagination and emotion.

As said earlier, the brilliant and awe-inspiring, creator, maker or progenitor of the current cosmos exists in two forms or versions, one at a time.

Its first form or version is its **original** form or version which is its **dimensionless** form or version. In this form or version, it existed some 13.7 billion light years ago only. So, it does not exist now in that form or version. It exists now in its second form or version which is described as its **3-D** or **three-dimensional** form or version.

Thus, the second form or version of the creator,

maker or progenitor of the current cosmos is its **3-D** or **three-dimensional** form or version. And this **3-D** or **three-dimensional** form or version   is its current form or version.

The current **3-D** or **three-dimensional** form or version of the creator, maker or progenitor of the present-day cosmos is described as the incredible, ubiquitous and infinite, **3-D** or **three-dimensional field of consciousness** or, **ocean of consciousness** called **cosmic space.**

The **original** form or version or, the **dimensionless** form or version of the creator, maker or progenitor of the current cosmos, which existed some 13.7 billion light years ago, is described as the incredible or awe-inspiring and **the only one of its kind** or unique, **timeless, bodiless,** and **dimensionless, consciousness** or **awareness** of infinite intelligence, imagination and emotion.

To recap.

The **human consciousnesses** are **timeless, immortal,** or **eternal** in the manner of their **fountainhead** or **source.** And this **timeless, immortal** or **eternal, fountainhead** or **source** of the **timeless, immoral,** or **eternal human consciousnesses** is the ubiquitous and infinite, **3-D** or **three-dimensional field of consciousness** or, **ocean of consciousness** called **cosmic space** which in turn is nothing but the expanded, distended, dilated or inflated form or version of the original, dimensionless form or version of the creator, maker or progenitor of the current cosmos.

The human consciousnesses are timeless, immortal or eternal in the manner of their timeless, immortal or eternal fountainhead or source because all human consciousnesses are a pure or pristine form or version of a segment of their timeless, immortal, or eternal, fountainhead or source. And their timeless, immortal, or eternal, fountainhead or source is the amazing, ubiquitous and infinite, 3-D or three-dimensional field of consciousness, or ocean of consciousness called, cosmic space.

The cosmic space, in turn, is nothing but the expanded, distended, dilated, or inflated, form or version of the original, dimensionless form or version of the incredible creator, maker or progenitor of the current cosmos.

The original form or version or the dimensionless form or version of the creator, maker or progenitor of the current cosmos is described as the incredible or awe-inspiring and the only one of its kind or unique, timeless, bodiless and dimensionless, consciousness or awareness of infinite intelligence, imagination and emotion.

The **original** form or version, or the **dimensionless** form or version of the creator, maker or progenitor of the current cosmos existed some 13.7 billion light years ago only and does not exist now.

As said before, the creator, maker or progenitor of the current cosmos exists now in its **3-D** or **three-dimensional** form or version which is none other than the amazing, ubiquitous and infinite, **3-D** or **three-dimensional, field of consciousness,** or **ocean of consciousness,** called **cosmic space.**

It occurs only to a very few **human consciousnesses** that their **true identity** consists of the **timeless** and **dimensionless consciousness** and **consciousness** only and nothing else. And the **physical body** inside which the **human consciousnesses** have been currently made to live or reside temporarily by their **fountainhead** or **source** is a mere short-term, internment, bondage or jail for them and nothing else, or if it is preferred, is a mere short-term digression or footnote for them and nothing else.

What has been said above can be put in another way.

Unfortunately, very few **human consciousnesses** accept the fact that their **physical bodies** are mere **transit camps** where they have been made to halt temporarily by their **fountainhead** or **source** so that they can discharge their respective **character role** or

performance role in the brilliantly produced, scripted and directed **cosmic play** or **drama** called the **physical cosmos** whose producer, script writer and director is none other than the selfsame or the very same **fountainhead** or **source** of the current cosmos.

Soon these **human consciousnesses** will depart from this **transit camp** called their **physical bodies** and merge back into their **fountainhead** or **source** namely, the amazing, ubiquitous and infinite **3-D** or **three-dimensional field of consciousness** or, **ocean of consciousness** called **cosmic space.**

The **human physical bodies** sans their **consciousnesses** are innately or inherently **insentient, insensate, inanimate** or **dead to the world** entities.

In total contrast, the **human consciousnesses** are innately, inherently or intrinsically **sentient** or **sensate** entities. Thy thus, are thinking, feeling, dreaming, perceiving, knowing, noticing, receptive, reactive, responsive, attentive, plus self-aware, self-cognizant or self-knowing entities.

Furthermore, the **human consciousnesses** are **non-physical** or **non-material** and thus, **non-dimensional** or **dimensionless truths** or, if it is preferred, are thus, **non-space-occupying truths.**

In comparison, the **human bodies -** inside which the

non-physical or **non-material** and thus, the **non-dimensional** or **dimensionless** or, **non-space-occupying, human consciousnesses** are made to live temporarily by their **fountainhead** or **source** - are **physical** or **material truths** or, if it is preferred, are **dimensional** or, **3-D** or **three-dimensional truths** or, better still, are **3-D** or **three-dimensional, space-occupying truths.**

To sum up.

The **human consciousnesses,** in total contrast to **human physical bodies,** are innately, inherently or intrinsically **cognizant, sentient** or **sensate** entities and thus, they also are innately, inherently or intrinsically, thinking, feeling, dreaming, perceiving, knowing, noticing, receptive, reactive, responsive, attentive, plus self-aware, self-cognizant or self-knowing entities.

The above is the case with **human consciousnesses** i.e. that they innately, inherently or intrinsically are **cognizant, sentient** or **sensate** entities due to the fact that they are a pure or pristine form or version of a segment of the amazing, ubiquitous and infinite, **3-D** or **three-dimensional field of consciousness** or, **ocean of consciousness** called **cosmic space** which in turn is nothing but the **expanded, distended, dilated** or **inflated** form or version of the **original** form or version or the **dimensionless** form or version of the incredible creator, maker or progenitor of the current cosmos.

In comparison, the **human physical bodies**, even though they also are nothing but a **segment** of the amazing, ubiquitous, and infinite, **3-D** or **three-dimensional, field of consciousness** or, **ocean of consciousness**, called **cosmic space**, in the manner **human consciousnesses** are, however, the **human physical bodies** are a **condensed, compressed** or **compacted** segment of the amazing, ubiquitous and infinite, **3-D** or **three-dimensional field of consciousness** or, **ocean of consciousness** called **cosmic space** and not a **pure** or **pristine** segment of the amazing, ubiquitous and infinite, **3-D** or **three-dimensional, field of consciousness**, or, **ocean of consciousness** called **cosmic space**, in the manner **human consciousnesses** are.

As a result, the **human physical bodies** temporarily lose their innate power of **self-awareness.** What's more, they also become **dimensional** or, **3-D** or **three-dimensional** in nature instead of remaining **dimensionless** or **non-dimensional** in nature which was their **original** nature when they were in their pure or pristine form and not in the condensed, compressed or compacted form as is the case with them now on account of them becoming the physical bodies of the human beings.

As a consequence of becoming **3-D** or **three-**

dimensional in nature, the **human physical bodies** come to be in the need of **space** or, better still, come to be in the need of **consciousnessbal space** for their **spatial placement** and **existence.** They thus, become burdened with the **tag** of being **'physical'** or **'material'.**

This **tag** of being **'physical'** or **'material'** will cling to **human bodies** till **'the very end of the current time',** meaning thereby, this **tag** of being **'physical'** or **'material'** will cling to the **human bodies** till **'the very demise or the very dissolution of the current cosmos'.** After **'the demise or the dissolution of the current cosmos'** or, after **'the demise or the dissolution of the current time',** the **human physical bodies** will become once again their **original-self** which is nothing but their **timeless, dimensionless** and **conscious-self.** That is to say, **after 'the demise or the dissolution of the current cosmos'** or, after **'the demise or the dissolution of the current time',** the **human physical bodies** will become once again, the **timeless, dimensionless consciousness,** and merge back into the **timeless, dimensionless consciousness** of the creator, maker or progenitor of the current cosmos.

Here a drop of ocean water which has become **transmuted** into a bit of solid ice on one hand and the ocean itself on the other, will be a very apt example in order to grasp the relationship that obtains between the ubiquitous and infinite, **3-D** or three-dimensional field of consciousness or, ocean of consciousness, called cosmic space on one hand and the human physical bodies on the other. Just as an ocean-drop, even after its metamorphosis into a bit of solid ice, is still, fundamentally or intrinsically, a small bit of ocean water and nothing else, likewise human bodies, despite possessing a gross, solid, 3-D or three-dimensional, physical, or material look, feel, behavior, demeanor or appearance, are nevertheless, fundamentally, intrinsically or inherently, nothing but a small bit of timeless, dimensionless, consciousness and nothing but a small bit of timeless, dimensionless consciousness, albeit in a metamorphosed or transmuted form.

What has been said above can be put in another way.

Just as a drop of ocean water even after its transmutation into a bit of solid ice, is nevertheless, fundamentally, intrinsically or inherently, nothing but a drop of ocean water, likewise human bodies, despite possessing a gross, solid, 3-D or three-dimensional, physical, or material look, feel, demeanor, behavior or appearance, are nevertheless, fundamentally, intrinsically or inherently, nothing but the metamorphosed form or version of a segment of the amazing, ubiquitous and infinite, 3-D or three-dimensional, field of consciousness or, **ocean of consciousness** called **cosmic space.**

Now let's return again to the subject of the creator, maker or progenitor of the current cosmos.

The incredible creator, maker or progenitor of the current cosmos, when it exists in its **original** form or version or, when it exists in its **dimensionless** form or version, it exists as an incredible or awe-inspiring and **the only one of its kind** or unique, **timeless, bodiless** and **dimensionless consciousness** or **awareness** of infinite intelligence, imagination and emotion.

Presently, however, this creator, maker or progenitor of the current cosmos exists in its **3-D** or **three-dimensional** form or version which is none other than the amazing, ubiquitous and infinite, **3-D** or **three-dimensional field of consciousness** or, **ocean of consciousness** called **cosmic space.**

The **human consciousnesses** rarely, if ever, comprehend the truth that they are not a **'perceivable physical object'** in the manner, **human physical bodies** are, the **human physical bodies** inside which the **human consciousnesses** are made to live or reside temporarily by their **fountainhead** or **source** namely, the amazing, ubiquitous and infinite **3-D** or **three-dimensional field of consciousness** or, **ocean of consciousness** called **cosmic space.**

Instead, the **human consciousnesses** are 'non-perceivable subjects' in the manner all **consciousnesses** are.

The expression or the word **'subject'** refers to **human consciousnesses** which, innately, inherently or intrinsically, are **non-perceivable, non-dimensional, non-physical** or **non-material** plus thinking, feeling, knowing, noticing, dreaming, perceiving, responsive, receptive, attentive and self-aware, self-cognizant, or self-knowing entities.

A **human consciousness,** as long as it clings to its delusion or wrong notion that its **identity** consists of the **time-bound, temporal,** or **mortal, human physical body,** so long it will be called an **'infra-transcendental consciousness'** or a **consciousness** which is yet **to transcend** or yet **to rise above** its long held and deeply rooted **delusion** or **wrong notion** that its **identity** consists of the **time-bound, temporal,** or **mortal, human physical body.**

In other words, as long as a **human consciousness** is firmly tied to the **'body consciousness'** and nothing but the **'body consciousness'** at the total expense or cost of, or, better still, at the total exclusion of **'consciousnessbal consciousness',** so long it will be called an **'infra-transcendental consciousness'** and nothing but an **'infra-transcendental consciousness '.**

The moment a **human consciousness** succeeds in getting rid of its long held and deeply rooted **'body consciousness'** or, succeeds in overcoming its long held and deeply rooted **delusion** that its **identity** consists of the **time-bound, temporal,** or **mortal,** plus **3-D** or **three-dimensional** and thus, **space-occupying, human physical body,** and instead, it establishes itself firmly into the **'consciousnessbal consciousness'** , meaning thereby that, it instead, establishes itself firmly into the **truth** that, in fact, its **identity** consists of the **timeless, dimensionless, non-material** and thus, **non-space occupying** plus **non-perceivable,** thinking, feeling, knowing, noticing, dreaming, perceiving, responsive, receptive, attentive and self-aware, self-cognizant, or self-knowing entity called the **consciousness** and nothing but the **consciousness,** at that very instant it becomes entitled to be called a **'transcendental consciousness'.**

Thus a **'transcendental consciousness'** is a **human consciousness** who has totally **transcended** its **'body consciousness'** and instead, has established itself firmly into the **'consciousnessbal consciousness'.**

What has been said above can be put in another way.

A **'transcendental consciousness'** is defined as a particular kind of **human consciousness** which has given up its long held and deep rooted delusion that

384

its **identity** consists of the **time-bound, temporal,** or **mortal, 3-D** or **three-dimensional** and thus **space-occupying human physical body** and has instead, entered **consciousnessbally** into a **higher state** where it correctly comprehends that, in fact, it's **true identity** consists of the **timeless, dimensionless, non-physical** and thus **non-space-occupying 'subject'** or, better still, it's **true identity** consists of the **timeless, dimensionless, non-physical** and thus **non-space-occupying** plus **non-perceivable** but **perceiving,** thinking, feeling, dreaming, knowing, noticing, enjoying, responsive, receptive, attentive and self-aware or self-knowing entity called the **consciousness** and nothing else.

~*~*~*~*~

ABOUT THE AUTHOR

Dr. Chandra Bhan Gupta, was born and educated in Lucknow, India.

He commenced his medical career in India with several notable medical articles to his credit.

Subsequently, he went to UK., where he continued his distinguished medical career, gaining the highest postgraduate and honorary accolades within his field.

Such questions as how and why man and the rest of creation have come into being, as well as the true nature of the creator and where is his abode, troubled him from an early age.

In an attempt to find answers to these eternal questions, he went through extreme austerities or penance over the course of many years, accompanied by long periods of deep meditation.

Enlightenment from the Almighty came in 1995, which resulted in the writing of the first book on the theme of "Supra-Spirituality", called '**Adwaita Rahasya: Secrets of Creation Revealed'**, followed by two more books delving deeper into the same theme, entitled **'Space is The Mind of God: A Scientific Explanation of God and His Abode'**, and the present work **'Cosmic Space is God & Physical Universe is God's Dream'**.

Dr. Chandra Bhan Gupta

Cosmic Space is God and Physical Universe is God's Dream

Printed in Poland
by Amazon Fulfillment
Poland Sp. z o.o., Wrocław

51796384R00237